Generis
PUBLISHING

Future Directions in Research on Skin Spottiness in Rainbow Trout

Integrating Basic and Applied Sciences

Nelson Colihueque Vargas

Title: **Future Directions in Research on Skin Spottiness in Rainbow Trout**

Integrating Basic and Applied Sciences

ISBN: 979-8-89966-681-0

Author: Nelson Colihueque Vargas

Cover image: www.pixabay.com

Publisher: Generis Publishing
Online orders: www.generis-publishing.com
Contact email: info@generis-publishing.com

Future Directions in Research on Skin Spottiness in Rainbow Trout: Integrating Basic and Applied Sciences

Nelson Colihueque Vargas
Departamento de Ciencias Biológicas y Biodiversidad
Universidad de Los Lagos
Osorno, Chile

CONTENTS

Preface

This book covers investigations that have been carried out on the skin spottiness of rainbow trout (*Oncorhynchus mykiss*), a typical skin pigmentation attribute of the species. Past and present advances in basic and applied research on this topic are discussed—in particular, the intraspecific variability of this trait and its degree of association with other skin pigmentation attributes, as well as with biometric and reproductive characteristics. This book highlights the use of computer vision frameworks to improve the accuracy of the skin spotting analysis. This computer-based image analysis offers more objective measurements than visual scoring estimates for a detailed quantification of the skin spotting, considering that the expression of this trait in rainbow trout is quite complex. This volume also highlights the commercial and biotechnological implications of the findings related to rainbow trout skin spotting.

The first chapter comments on the general characteristics of the skin pigmentation of rainbow trout, the cellular basis for its origin, and the main skin spotting characteristic of the species.

The second chapter describes the inter-individual variation in the level of spotting at the juvenile stage based on the investigation by Colihueque (2010b), focusing on understanding the ontogeny of skin spotting that determines its variability at advanced development stages.

The third chapter analyses the skin spotting variability of naturalised rainbow trout populations based on the research by Colihueque et al. (2018). It sheds light on how this trait is associated with biometric and reproductive parameters.

The fourth chapter comments on the growth performance of blue-coloured rainbow trout reared under intensive farming, based on the study by Colihueque and Estay (2018). This phenotype has a favourable external appearance for marketing, especially when it exhibits less spotted skin; therefore, specific analyses are required to clarify its growth rate related to the level of skin spotting. This research has provided favourable evidence of better growth performance when fish exhibit a highly intense blue back colour and a negative correlation between blue dorsal colour and the level of skin spotting.

The fifth chapter discusses a novel biotechnological approach based on the study by Colihueque and Estay (2025), which provides evidence that favours the application of the flank skin spotting pattern for the photo-identification of rainbow trout individuals.

This method could be used as an alternative non-invasive tagging method for addressing different biological questions related to this species.

Chapter 1.

The skin pigmentation of rainbow trout

1. General characteristics of the skin pigmentation

The Salmonidae family is known for its great variation in skin pigmentation and pigment patterning at both the inter- and intraspecific levels (Ade 1989). However, the mechanisms leading to such variation, including the pigment cells involved and their interactions, are mostly unknown (Djurdjevič et al. 2015). Among salmonids, rainbow trout (*Oncorhynchus mykiss*) shows distinctive skin pigmentation, characterised by a light background colour; dark spots, usually numerous on the back and the dorsal or caudal fins; and a red or purple band located across the flanks of the fish, including the gill covers (Ade 1989). This skin pigmentation is well established at the adult stage and is representative of the wild-type phenotype of the species (Fig. 1). However, the skin pigmentation of this species can vary greatly within and between populations. For example, in cultured populations of rainbow trout, the intraspecific variability affects the back, which can appear blue, yellow, or green in colour, and the flank and back, where dark spots may vary in number and size (Islam et al. 1973; Tack 1973; Colihueque et al. 2011, 2014, 2018). Even the skin on the back and belly may vary in its level of darkness, resulting in individuals that vary from light to dark (Colihueque et al. 2011). Notably, rainbow trout distributed throughout western North America (i.e., in its native range) also shows skin pigmentation variations. The two evolutionary lines coexist in this geographical area, which are composed of coastal rainbow trout and interior redband trout, show dissimilar skin pigmentation (Behnke 1979). Coastal rainbow trout has heavily spotted skin both above and below the lateral line with irregularly shaped spots, while interior redband trout presents relatively small spots, mainly restricted above the lateral line and on the caudal peduncle. In addition, the skin pigmentation of rainbow trout can also change as function of the developmental and maturation stages (Ade 1989). For example, young fish, known as parr, present characteristic small oval and rounded parr marks on the body, and small dark spots on the dorsal fin and tail. In contrast, mature adult fish usually show a pink or red stripe along the flanks, and the caudal fin is fully spotted.

2. Cellular basis of the skin pigmentation

Melanophores, xanthophores and iridophores—collectively referred to as chromatophores—are the cells responsible for generating the skin pigmentation in rainbow trout (Harris and Hunt 1973; Hawkes 1974; Yamaguchi and Miki 1981; Oguri 1983; Hattori et al. 2020) (Table 1). They are embryologically derived from neural crest cells (Erickson et al. 1998) and can exhibit a characteristic colour, given their ability to store or synthesise a specific pigment. Melanophores generate black or brown colours because they contain the light-absorbing pigment known as melanin, xanthophores give rise to yellow or orange colours due to their pteridines or carotenoids, and iridophores yield iridescent or silver colours because they contain light-reflecting platelets of guanine.

Chromatophores are located within the dermis and epidermis of the skin, where they generate colour patterns through their combined and differential spatial distribution. In rainbow trout, melanophores are located beneath the epidermis, in proximity to the scales, as well as in deeper layers of the dermis, as shown by histological observations (Oguri 1983) (Fig. 2). Light and electron microscopy analyses of coho salmon skin further indicate that chromatophores may occur either as isolated or as multicellular assemblies. In the latter configuration, melanophores and iridophores are closely associated, forming structures known as chromatophore units (CUs) (Hawkes 1974). Xanthophores are distributed in a more random manner, either above or below these units. This organisation produces a range of colour tones across different regions of the skin. In particular, the abundance of iridophores in specific areas, such as ventral skin, contributes to a characteristic silvery appearance.

Both melanophores and xanthophores exhibit motility, as they are capable of transitioning from punctuate (aggregated) state to a dendritic (dispersed) state, and vice versa, in response to environmental stimuli such as light intensity or exposure to certain chemical solutions (Hawkes 1974). For example, under increased light intensity, or following treatment of melanophores with 0.5% NaCl, skin darkening occurs as melanosomes are transported towards the cell periphery, thereby increasing light absorption (Fig. 2). Conversely, exposure to 0.5% KCl induces pigment aggregation, resulting in a lighter appearance. At the physiological level, these rapid changes are regulated by two hormones secreted by the pituitary: α-melanophore-stimulating hormone (MSH), which promotes dispersion of pigment granules, and melanophore-concentrating hormone (MCH), which exerts an antagonist effect by inducing pigment aggregation (Oshima et al. 2001). Collectively, these mechanisms underpin physiological colour change (Fujii 2000), enabling fish to rapidly adjust their skin pigmentation to match environmental background conditions, whether light or dark.

3. Skin spottiness in rainbow trout

Spottiness is a distinctive feature of skin pigmentation in rainbow trout. In adult individuals, this trait consists of conspicuous small dark spots mainly distributed on the flank (above and below the lateral line) and on the back, including the operculum and the dorsal and caudal fins (Ade 1989). However, spotting patterns exhibit considerable variation in terms of number, size, and position on the body. Islam et al. (1973) and Tack (1973) were among the first researchers to document such variations in cultured stocks of rainbow trout from Japan and Germany, respectively, based on the quantification of the number of dark spots below the lateral line (Fig. 3). Subsequent studies have further confirmed the occurrence of this phenomenon in this species (Kause et al. 2003; Colihueque 2010; Colihueque et al. 2011, 2014, 2018; Rodríguez et al. 2019). For example, it has been observed that there is intraspecific variation in the number of spots below the lateral line (Islam et al. 1973; Tack 1973; Colihueque et al. 2018), above the lateral line (Colihueque and Estay 2018), and on the back (Colihueque et al. 2011, 2014), as well as variation in the size of spots on the flanks (Kause et al. 2003) and above the lateral line (Colihueque and Estay 2018) (Fig. 4). Accordingly, individuals may range from lightly spotted or nearly unspotted phenotypes to highly spotted phenotypes across body regions, as observed in cultured populations from Chile (Fig. 5). Quantitative analyses indicate that the spot density is higher on the dorsal region than on the flank (mean ± SD (standard deviation): 2.7 ± 0.1 vs 1.2 ± 0.1 spots/cm^2), with a normal distribution observed dorsally and an exponential distribution on the flank (Colihueque et al. 2011). The arrangement of spots in the skin (i.e., regularity) is another feature that can show variability in rainbow trout. In the back region of fish, it can show variation between blue-back and non-blue-back phenotypes because blue-back individuals display a significantly greater irregularity in dark spot distribution (Colihueque 2014). This irregularity can also be observed in the spot disposition in the flank of individuals (Colihueque and Estay 2025).

To date, the origin of this variation remains unclear; however, the number of spots may covary with fish size and with biometric and reproductive traits for spots located below the lateral line (Islam et al. 1973; Colihueque and Estay 2018), as well as with age for dorsal spotting (Colihueque et al. 2011). Additional evidence suggests that variation in skin spotting is under genetic control, as indicated by medium to low heritability estimates ($h^2 = 0.16–0.45$) for the number of dark spots on the dorsal surface (Díaz et al. 2011) and above and below the lateral line (Rodriguez et al. 2019), as well as the size of the spots on the flanks (Kause et al. 2003). These findings support significant polygenic determination underlying this trait.

4. Concluding remarks

Spottiness is a distinctive feature of skin pigmentation in rainbow trout, which typically exhibits substantial intraspecific variability. This trait is associated with the presence of isolated or clustered melanophores located in the epidermis or dermis. Previous studies have begun to elucidate its relationships with biometric and reproductive parameters, its establishment during development, and its underlying genetic control. However, the mechanisms governing variations in spotting patterns in rainbow trout are still poorly understood and require further investigation.

5. References

Ade R (1989) Trout and salmon handbook. Facts on File, Inc., New York.

Behnke RJ (1979) The native trout of the genus *Salmo* of Western North America. USDA Forest Service, Fish and Wild Life Service, Bureau of Land Management, Lakewood, Colorado, 163 pp.

Colihueque N (2014) Analysis of the coloration and spottiness of Blue Back rainbow trout at a juvenile stage. Journal of Applied Animal Research 42: 474–480. https://doi.org/10.1080/09712119.2014.883317

Colihueque N (2010) Genetics of salmonid skin pigmentation: Clues and prospects for improving the external appearance of farmed salmonids. Reviews in Fish Biology and Fisheries 20: 71–86. https://doi.org/10.1007/s11160-009-9121-6

Colihueque N, Parraguez M, Estay FJ, Diaz NF (2011) Skin color characterization in rainbow trout by use of computer-based image analysis. North American Journal of Aquaculture 73: 249–258. https://doi.org/10.1080/15222055.2011.581578

Colihueque N, Estay FJ (2018) Association between the blue color intensity of the skin with growth performance and skin spottiness in a blue-colored strain of rainbow trout (*Oncorhynchus mykiss*). The Open Agriculture Journal 12: 251–261. https://doi.org/10.2174/1874331501819010251

Colihueque N, Estay FJ, Yáñez M (2018) Skin spotting variation associated with biometric and reproductive parameters in naturalized populations of rainbow trout, *Oncorhynchus mykiss*, from southern Chile. The Open Ecology Journal 11: 25–37. https://doi.org/10.2174/1874213001811010025

Colihueque N, Estay F (2025) Photo-identification of individual rainbow trout, *Oncorhynchus mykiss* (Actinopterygii, Salmoniformes, Salmonidae), using the flank spot pattern. Acta Ichthyologica et Piscatoria 55: 181–190. https://doi.org/10.3897/aiep.55.151044

Díaz NF, Estay FJ, Araneda C, Colihueque N (2011) Selección genética de rasgos del color de la piel de la trucha arcoiris *Oncorhynchus mykiss*. Versión diferente Enero 2011: 56–57.

Djurdjevič I, Kreft ME, Sušnik Bajec S (2015) Comparison of pigment cell ultrastructure and organisation in the dermis of marble trout and brown trout, and first description of erythrophore ultrastructure in salmonids. Journal of Anatomy 227: 583–595. https://doi.org/https://doi.org/10.1111/joa.12373

Erickson CA, Reedy M V (1998) Neural crest development: the interplay between morphogenesis and cell differentiation. In: Pedersen RA, Schatten GPBT-CT in DB (Eds), Academic Press, 177–209. https://doi.org/https://doi.org/10.1016/S0070-2153(08)60367-1

Fujii R (2000) The regulation of motile activity in fish chromatophores. Pigment Cell Research 13: 300–319. https://doi.org/10.1034/j.1600-0749.2000.130502.x

Harris JE, Hunt S (1973) The fine structure of iridophores in the skin of Atlantic salmon (*Salmo salar* L.). Tissue and Cells 5: 479–488.

Hattori RS, Yoshinaga TT, Butzge AJ, Hattori-Ihara S, Tsukamoto RY, Takahashi NS, Tabata YA (2020) Generation of a white-albino phenotype from cobalt blue and yellow-albino rainbow trout (*Oncorhynchus mykiss*): Inheritance pattern and chromatophores analysis. PLOS ONE 15: e0214034. https://doi.org/10.1371/journal.pone.0214034

Hawkes JW (1974) The structure of fish Skin II. The chromatophore unit. Cell Tissue Research 149: 159–172.

Islam A, Nose Y, Yasuda F (1973) Number and distribution pattern of the black spot on body surface of rainbow trout. Bulletin of the Japanese Society of Scientific Fisheries 39: 727–739.

Kause A, Ritola O, Paananen T, Eskelinen U, Mäntysaari E (2003) Big and beautiful? Quantitative genetic parameters for appearance of large rainbow trout. Journal of Fish Biology 62: 610–622. https://doi.org/10.1046/j.1095-8649.2003.00051.x

Oguri M (1983) Dermal melanophores and renal melanin-containing cells in the cobalt variant of rainbow trout. Nippon Suisan Gakkaishi 49: 1679–1681. https://doi.org/10.2331/suisan.49.1679

Oshima N, Nakamaru N, Araki S, Sugimoto M (2001) Comparative analyses of the pigment-aggregating and -dispersing actions of MCH on fish chromatophores. Comparative Biochemistry and Physiology Part C: Toxicology & Pharmacology 129: 75–84. https://doi.org/10.1016/S1532-0456(01)00187-9

Rodríguez FH, Cáceres G, Lhorente JP, Newman S, Bangera R, Tadich T, Neira R, Yáñez JM (2019) Genetic (co)variation in skin pigmentation patterns and growth in rainbow trout. Animal 13: 675–682. https://doi.org/10.1017/S175173111800188X

Tack E (1973) Züchtungsfragen in der forellenzucht. Österreichs Fischerei 26: 77–84. Available from: https://www.zobodat.at/publikation_articles.php?id=239216.

Yamaguchi K, Miki W (1981) Comparison of pigments in the integument of cobalt, albino, and normal rainbow trout, *Salmo gairdnerii irideus*. Comparative Biochemistry and Physiology Part B: Comparative Biochemistry 68: 517–520. https://doi.org/10.1016/0305-0491(81)90068-7

Table 1. Types of chromatophores and associated pigment and colour in rainbow trout.

Chromatophore	Organelle/structure	Pigment	Colour	Principle of coloration	Reference
Melanophore	Melanosomes	Melanin	Black, brown	Light absorption	Oguri (1983), Hattori et al. (2020)
Xantophore	Xanthosome	Pteridines, carotenoides	Yellow, orange	Absorption, reflection	Hattori et al. 2020
Iridophore	Reflecting platelets	Purines, guanine platelet	Structural colours or Tyndall effect, iridescence	Light reflection	Harris and Hunt (1973), Hawkes (1974)

Fig. 1 Wild-type phenotype of adult rainbow trout (total body length = 40 cm).

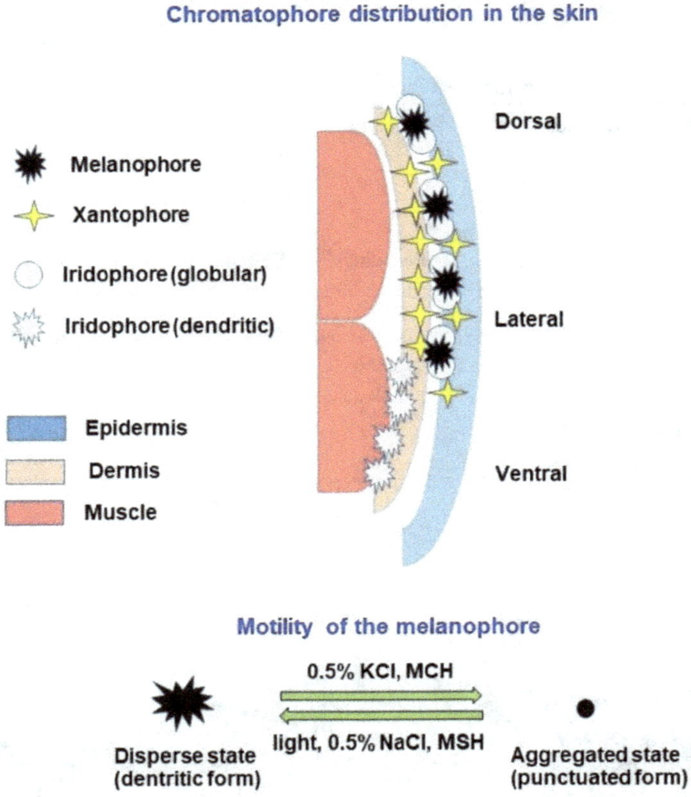

Fig. 2 Chromatophore distribution in the skin of adult rainbow trout. The motility of melanophores and the factors involved are also shown. MCH: melanophore-concentrating hormone; MSH: α-melanophore-stimulating hormone. The data are derived from Hawkes (1974) and Oguri (1983).

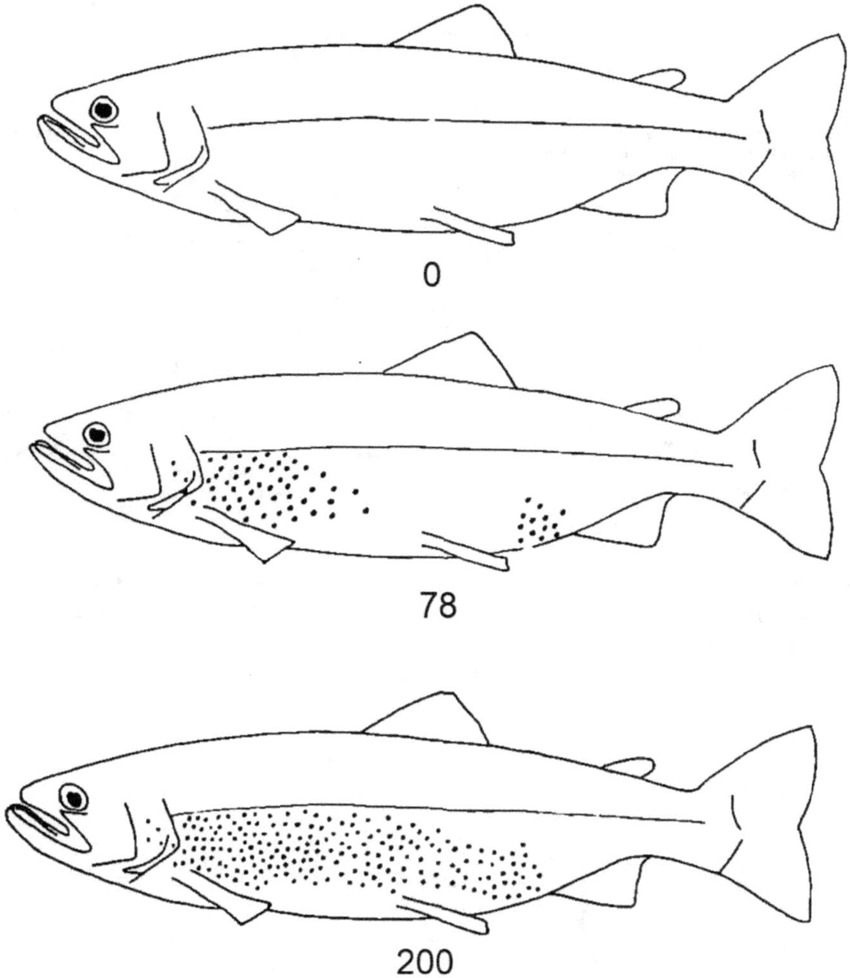

Fig. 3 Variation in the number of dark spots below the lateral line observed in a normal commercial rainbow trout population in a fish farm in Albaumer, Germany. The minimum (0), mean (78) and maximum (200) numbers of dark spots are shown. Source: Tack (1973).

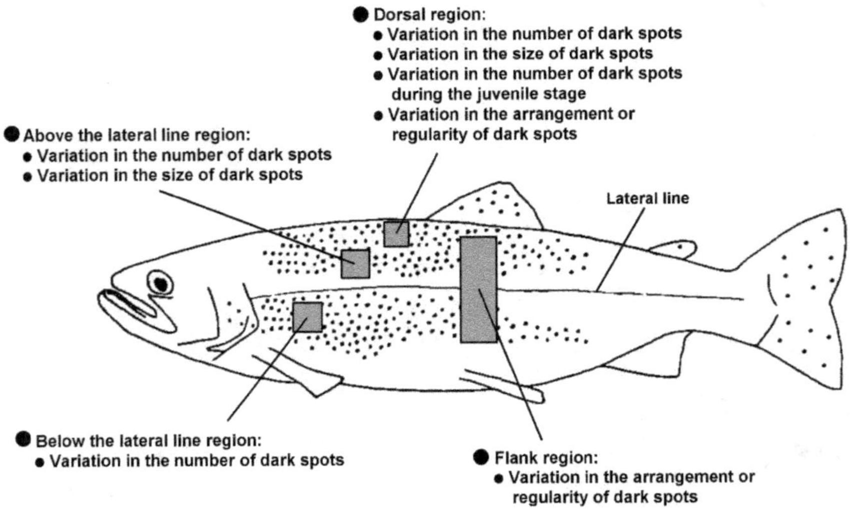

Fig. 4 Variation of skin spottiness observed in different body regions of rainbow trout. The data are derived from Islam et al. (1973), Tack (1973), (Kause et al. (2003), Colihueque et al. (2011), Colihueque et al. (2014), Colihueque et al. (2018), Colihueque and Estay (2018), Colihueque and Estay (2025).

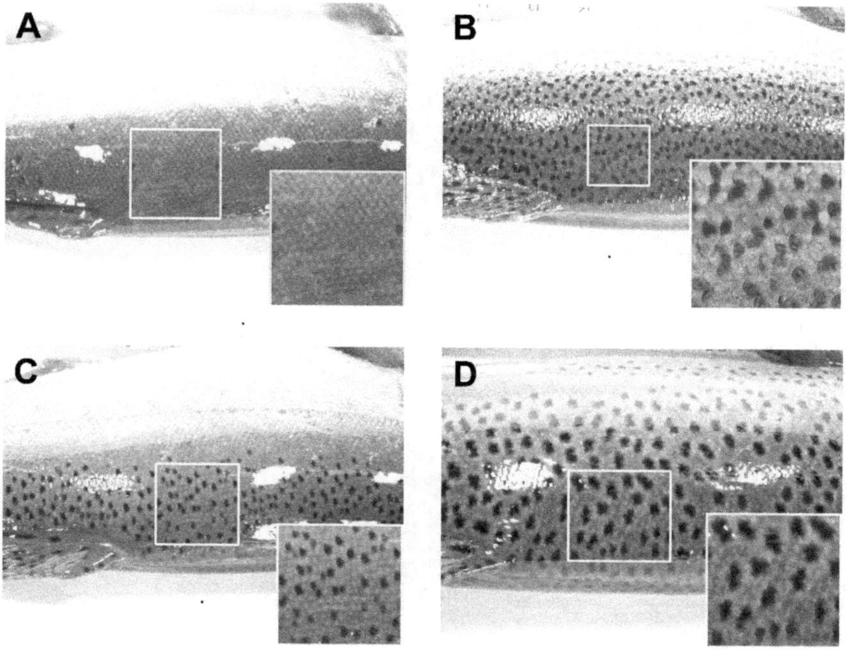

***Fig. 5** Variation in skin spotting on the back of adult rainbow trout. (A) and (B) show variation in the number of dark spots (0.2 and 4.4 dark spots/cm^2, respectively), while (C) and (D) show variation in the size of dark spots (0.04 and 0.12 cm^2, respectively). All individuals are 3-year-old female breeders.*

Chapter 2.

Inter-individual variation of skin spottiness in rainbow trout

1. Introduction

Characterising skin pigmentation during development can provide valuable data for understanding pigment pattern formation in fish. This analysis can greatly facilitate the identification of intraspecific variation in skin pigmentation, as well as the change or maintenance of specific phenotypes at different developmental stages. This task is required in rainbow trout because, to date, the establishment of different skin pigmentation attributes across body areas (fins, head, above and below the lateral line, etc.) and the origin of inter-individual variations are unclear. A more comprehensive understanding of this process may help explain the observed differences in the skin spottiness characteristics at the adult stage, particularly regarding the number and size of the dark spots. In contrast to the substantial advances in fish models, which have, for example, clarified the mechanisms underlying striped pattern formation in zebrafish (Kelsh 2000, 2004; Parichy 2003; Colihueque 2010a), research on this topic in rainbow trout remains limited. Studies in zebrafish have identified several key factors involved pigmentation pattern formation, including interactions among neighbouring chromatophores, the *de novo* generation of chromatophores, and the influence of the local cellular environment. To date, the only investigation addressing this issue in rainbow trout was published by our group in 2014. That study employed computer-based image analysis to examine changes in skin spottiness during juvenile development in rainbow trout classified into two groups according to back coloration, namely blue-back and non-blue-back phenotypes (Colihueque 2014).

It should be noted that there several studies have addressed this issue in rainbow trout; however, these investigations have been limited to the adult stage and have involved cultured populations from Japan (Islam et al. 1973) and Germany (Tack 1973), as well as naturalised populations from Chile (Colihueque et al. 2018). Collectively, these studies have provided evidence of significant associations between the number of dark spots and biometric traits. For example, Colihueque et al. (2018) reported a negative correlation between the number of dark spots below the lateral line and body weight in males and a positive correlation between the number of dark spots below the lateral line and body length in females. These results are consistent with previous investigations demonstrating a positive association between the number of dark spots below the lateral line and body length in this species, thereby suggesting a close relationship between skin spotting patterns and growth (Islam et al. 1973). Furthermore, a quantitative genetic analysis evaluating spotting intensity above and below the lateral line in cultured Chilean rainbow trout populations found significant low-to-moderate positive genetic correlations with harvest body weight. These findings support the notion that the level of spotting in this species are influenced by additive genetic effects (Rodriguez al. 2019).

2. Studies on the inter-individual variation in skin spottiness

Colihueque (2010b) performed an inter-individual analysis of juvenile rainbow trout during early developmental stages and provided relevant information on the origin of variation in skin spottiness variation in rainbow trout. The author analysed a cohort of cultured juvenile rainbow trout (n = 24) aged 282–447 days post fertilisation (dpf), obtained from the fish farm Huililco Ltda., Pucón district, La Araucanía Region, Chile. The specimens belonged to an F1 progeny derived from crosses between Steelhead females and Aquagen neomales. Fish were individually tagged with passive integrated transponders (PIT) at a body weight of 5.0 g to allow unambiguous identification.

Skin spottiness was analysed using a computer vision approach (Fig. 1), based on photographic records of each individual taken at nine developmental stages between June and November 2007. Whole-body images were captured in dorsal view using a 7.0-megapixel digital camera under controlled lighting conditions. The developmental stages corresponded to 282 (stage 1), 303 (stage 2), 318 (stage 3), 330 (stage 4), 365 (stage 5), 380 (stage 6), 400 (stage 7), 426 (stage 8), and 447 (stage 9) dpf (with accumulated thermal units [ATU = days * °C above 0°C] of 2538, 2727, 2934, 3096, 3285, 3465, 3663, 3834, and 4023, respectively). At stage 1, the mean body weight and total length were 37.5 ± 9.5 g and 14.5 ± 1.3 cm, respectively, increasing to 175.5 ± 42.3 g and 23.7 ± 2.4 cm at stage 9. For image analysis, a 1.5 × 1.5 cm region anterior to the dorsal fin was selected from each photograph, and spot characteristics were quantified using the particle analysis tool in ImageJ v1.38. Stage 1 (282 dpf) was selected as the starting point because dorsal spots first became distinguishable at this stage; earlier stages were characterised by parr marks consisting of small oval and rounded markings on the flanks. Spot density was quantified in grayscale and is presented as the number of dark spots per unit area (spots/cm^2), and the size of is expressed as the proportion of area occupied by dark markings per cm^2.

Across all stages, the mean number of dark spots was 1.80 ± 0.15 and the mean size of dark spots was 0.92 ± 0.10 (Table 1). Except for the first two stages, both variables followed a normal distribution according to the Kolmogorov–Smirnov test. Both the number and size of spots increased over the course of juvenile development, ranging from 0.74 ± 0.24 to 1.72 ± 0.42 for the number of spots and from 0.13 ± 0.05 to 1.55 ± 0.40 for the size of spots. However, the magnitude of change was substantially greater for the size of spots (11.9-fold increase) than for the number of spots (2.3-fold increase). This pattern was supported by linear Pearson correlation analysis, which showed a stronger association with age for the size of spots than for the number of spots ($r = 0.95$ and $r = 0.63$, respectively). These results indicate that developmental stage is more strongly associated with changes in the size of spots than the number of spots.

At the individual level, most fish exhibited negligible spotting at stage 1. Between stages 1 and 9, the number of spots increased in 14 individuals (58.3%), decreased in 6 (25%), and remained unchanged in 4 (16.6%) (Table 2). The positive Pearson correlation coefficients (r) between number of spots and age observed in most individuals (18/24) were consistent with an overall increase in spotting density during development. Some individuals showed a pronounced increase, reaching more than 2.22 spots/cm^2 between stages 1 and 9. However, this increase appeared to peak at approximately 330 dpf (3,096 ATU), as indicated by trend analysis (Fig. 2A). In addition, a subset of individuals remained entirely unspotted throughout development, resulting in spot-free dorsal skin. These patterns contributed to substantial inter-individual variation at later developmental stages, as illustrated by photographs of individuals at 426 dpf (Fig. 3).

There was a similar pattern for the size of spots, which increased in 18 individuals (75%), decreased in 2 (8.3%), and remained the same in 4 (16.6%) (Table 3). Most individuals showed a positive correlation between the size of spots and age (22/24), indicating a general increase in the size of spots during development. Among individuals exhibiting an increase in the size of spots, one-third (33.3%) showed marked increases, ranging from 2.13% to 8.80% (Fig. 2B).

3. Discussion

The results of this research indicate that dorsal skin spotting in juvenile rainbow trout aged 282–447 dpf is in the process of establishment. This establishment involves an increase in the number and size of spots, starting from zero or near-zero values for both parameters, with significant increases of 2.3 and 11.9 times, respectively. This increase is most pronounced during stages 1–4 (282–330 dpf), after which the skin spotting remain more constant, especially in terms of their number. In other words, there appears to be a specific developmental stage (330 dpf; 3,096 ATU) in which the dorsal skin spotting becomes uniform. However, not all individuals followed a similar skin spotting pattern because some individuals without dark spots at the initial stage did not develop new dark spots across stages, resulting in fish without any dark spots on their dorsal skin at stage 9.

These results indicate a differential appearance or lack of dark spots during the early development of rainbow trout, would explain the inter-individual variation in dorsal skin spotting in later developmental stages. This ontogenetic process gives rise to a wide skin spotting variation related to a change in the number and size of dark spots, whose occurrence has not been documented in previous publications. A study conducted by our research group demonstrated that rainbow trout exhibit ontogenetic variation in skin pigmentation patterns, particularly involving dorsolateral shifts in the position of dark spots during early development towards more dorsal regions in both blue-back and non-blue-back phenotypes (Colihueque 2014). Blue-back individuals were also found to display a significantly greater irregularity in dark spot distribution patterns. Furthermore, the dorsal skin spotting pattern established during early developmental stages may persist into adulthood at the individual level. This possibility is supported by evidence provided by our group demonstrating the conservation of flank skin spotting patterns across adult developmental stages, specifically between the ages of 2 and 3 years in rainbow trout (Colihueque and Estay 2025).

Changes in dark skin spotting in rainbow trout have been extensively documented in the scientific literature. Evidence indicates the existence of intraspecific variation in the number of spots below the lateral line (Islam et al. 1973; Tack 1973; Colihueque et al. 2018), above the lateral line (Colihueque and Estay 2018), and on the back (Colihueque et al. 2011, 2014), as well as variation in the size of spots on the flanks (Kause et al. 2003) and above the lateral line (Colihueque and Estay 2018). These traits typically exhibit considerable variability. For example, in cultured populations, the coefficients of variation for the number and size of dark spots above the lateral line range from 39% to 55% and from 46% to 53%, respectively (Colihueque and Estay 2018b). Consequently, it is common to observe individuals displaying extreme differences in skin spottiness, ranging from complete absence of dark spots to approximately eight dark spots per square centimetre above the lateral line (Colihueque and Estay 2018). Despite the extensive phenotypic variation described in this species, the underlying causes of skin spottiness variability remain largely unknown. Nevertheless, available evidence suggests that this variation may be associated with biometric and reproductive traits (Islam et al. 1973; Colihueque et al. 2018), as well as with the age of the individuals (Colihueque et al. 2014).

The available evidence suggests that quantitative genetic factors are likely involved in controlling the level of skin spotting in rainbow trout (Kause et al. 2003; Díaz et al.

2011; Rodríguez et al. 2019). This genetic influence may be mediated through quantitative trait loci (QTLs) regulating phenotypic variation in this trait. Although such QTLs have not yet been identified in rainbow trout, studies in the threespine stickleback fish model have already reported QTLs associated with the regulation of melanophore number in the skin (Greenwood et al. 2011). In addition to genetic influences, the potential contribution of environmental factors to skin spotting variation in rainbow trout cannot be excluded. In other salmonid species, including brown trout (Blanc et al. 1982) and grayling (Penteriani et al. 2015), skin spotting levels have been shown to be significantly affected by environmental conditions. Furthermore, the involvement of specific genes controlling skin spottiness in rainbow trout should also be considered in light of advances achieved in fish model systems. For instance, in zebrafish, several genes associated with pigmentation patterning have been identified. Among these, mutations in the *connexin41.8* gene produce the leopard phenotype, characterised by pronounced variation in spot patterning (Watanabe et al. 2006). Similarly, the *kit* and *ednrb1* genes have been demonstrated to regulate the development of the normal complement of adult melanophores in this species (Parichy et al. 1999, 2000). Consequently, future molecular-genetic investigations targeting genes of this class may provide valuable insights into the mechanisms underlying skin spotting variation in rainbow trout.

4. Conclusion

Dorsal skin spotting in rainbow trout is established during the juvenile stage between 282 and 447 dpf. This developmental process involves progressive increases in the number and size of spots. A distinctive developmental stage (330 dpf; 3,096 ATU) appears to mark the point at which dorsal spotting stabilises in terms of the number of spots. However, some individuals did not develop dorsal spotting during this period, resulting in an unspotted phenotype. Overall, these ontogenetic dynamics likely contribute to inter-individual variation in dorsal spotting observed at later developmental stages and may help explain the phenotypic variability commonly reported in both cultured and naturalised populations of this species.

5. References

Blanc JM, Poisson H, Vibert R (1982) Variabilité génétique de la ponctuation noire sur la truitelle Fario (*Salmo trutta* L.). Annales de génétique et de sélection animale 14: 225. https://doi.org/10.1186/1297-9686-14-2-225

Colihueque N (2014) Analysis of the coloration and spottiness of Blue Back rainbow trout at a juvenile stage. Journal of Applied Animal Research 42: 474–480. https://doi.org/10.1080/09712119.2014.883317

Colihueque N (2010a) Genetics of salmonid skin pigmentation: Clues and prospects for improving the external appearance of farmed salmonids. Reviews in Fish Biology and Fisheries 20: 71–86. https://doi.org/10.1007/s11160-009-9121-6

Colihueque N (2010b) Caracterización fenotípica y análisis genético-molecular de la pigmentación de la piel en trucha arcoíris (*Oncorhynchus mykiss*). Tesis de doctorado en Ciencias Silvoagropecuarias y Veterinarias, Universidad de Chile

Colihueque N, Parraguez M, Estay FJ, Diaz NF (2011) Skin color characterization in rainbow trout by use of computer-based image analysis. North American Journal of Aquaculture 73: 249–258. https://doi.org/10.1080/15222055.2011.581578

Colihueque N, Estay F (2025) Photo-identification of individual rainbow trout, *Oncorhynchus mykiss* (Actinopterygii, Salmoniformes, Salmonidae), using the flank spot pattern. Acta Ichthyologica et Piscatoria 55: 181–190. https://doi.org/10.3897/aiep.55.151044

Colihueque N, Estay FJ (2018) Association between the blue color intensity of the skin with growth performance and skin spottiness in a blue-colored strain of rainbow trout (*Oncorhynchus mykiss*). The Open Agriculture Journal 12: 251–261. https://doi.org/10.2174/1874331501819010251

Colihueque N, Estay FJ, Yáñez M (2018) Skin spotting variation associated with biometric and reproductive parameters in naturalized populations of rainbow trout, *Oncorhynchus mykiss*, from southern Chile. The Open Ecology Journal 11: 25–37. https://doi.org/10.2174/1874213001811010025

Díaz NF, Estay FJ, Araneda C, Colihueque N (2011) Selección genética de rasgos del color de la piel de la trucha arcoiris *Oncorhynchus mykiss*. Versión diferente Enero 2011: 56–57.

Greenwood AK, Jones FC, Chan YF, Brady SD, Absher DM, Grimwood J, Schmutz J, Myers RM, Kingsley DM, Peichel CL (2011) The genetic basis of divergent pigment patterns in juvenile three spine sticklebacks. Heredity 107: 155–166. https://doi.org/10.1038/hdy.2011.1

Islam A, Nose Y, Yasuda F (1973) Number and distribution pattern of the black spot on body surface of rainbow trout. Bulletin of the Japanese Society of Scientific Fisheries 39: 727–739.

Kause A, Ritola O, Paananen T, Eskelinen U, Mäntysaari E (2003) Big and beautiful? Quantitative genetic parameters for appearance of large rainbow trout. Journal of Fish Biology 62: 610–622. https://doi.org/10.1046/j.1095-8649.2003.00051.x

Kelsh RN (2004) Genetics and evolution of pigment patterns in fish. Pigment Cell Research 17: 326–336. https://doi.org/10.1111/j.1600-0749.2004.00174.x

Kelsh RN, Schmid B, Eisen JS (2000) Genetic analysis of melanophore development in zebrafish embryos. Developmental Biology 225: 277–293. https://doi.org/10.1006/dbio.2000.9840

Parichy DM (2003) Pigment patterns: fish in stripes and spots. Current Biology 13: R947–R950. https://doi.org/10.1016/j.cub.2003.11.038

Parichy DM, Mellgren EM, Rawls JF, Lopes SS, Kelsh RN, Johnson SL (2000) Mutational analysis of endothelin receptor b1 (rose) during neural crest and pigment pattern development in the zebrafish *Danio rerio*. Developmental Biology 227: 294–306. https://doi.org/10.1006/dbio.2000.9899

Parichy DM, Rawls JF, Pratt SJ, Whitfield TT, Johnson SL (1999) Zebrafish sparse corresponds to an orthologue of *c-ki*t and is required for the morphogenesis of a subpopulation of melanocytes, but is not essential for hematopoiesis or primordial germ cell development. Development 126: 3425–3436. https://doi.org/10.1242/dev.126.15.3425

Penteriani V, Delgado MM, Clavero M, Ivanc M (2015) Spottiness of European graylings (Actinopterygii, Salmonidae) correlates with environmental features and could signal the status of individuals. Italian Journal of Zoology 82: 202–211. https://doi.org/10.1080/11250003.2015.1018852

Rodríguez FH, Cáceres G, Lhorente JP, Newman S, Bangera R, Tadich T, Neira R, Yáñez JM (2019) Genetic (co)variation in skin pigmentation patterns and growth in rainbow trout. Animal 13: 675–682. https://doi.org/10.1017/S175173111800188X

Tack E (1973) Züchtungsfragen in der forellenzucht. Österreichs Fischerei 26: 77–84. Available from: https://www.zobodat.at/publikation_articles.php?id=239216.

Watanabe M, Iwashita M, Ishii M, Kurachi Y, Kawakami A, Kondo S, Okada N (2006) Spot pattern of leopard *Danio* is caused by mutation in the zebrafish *connexin41.8* gene. The EMBO Reports 7: 893–897. https://doi.org/10.1038/sj.embor.7400757

Table 1. Mean values of the skin spottiness on the back of juvenile rainbow trout at different developmental stages. Mean ± SE (standard error) is shown. Source: Colihueque (2010b)

Developmental stage	Age (days post-fertilisation)	ATU[&]	Sample size	Spotting density (Number of dark spots/cm^2)	Size of spots (Percentage of dark spots/cm^2)
1	282	2,538	24	0.74 ± 0.2*	0.13 ± 0.05*
2	303	2,727	24	0.89 ± 0.23*	0.21 ± 0.06*
3	318	2,934	24	1.81 ± 0.53^{NS}	0.56 ± 0.19^{NS}
4	330	3,096	24	2.17 ± 0.54^{NS}	0.80 ± 0.26^{NS}
5	365	3,285	24	2.15 ± 0.45^{NS}	0.99 ± 0.24^{NS}
6	380	3,465	24	2.32 ± 0.45^{NS}	1.18 ± 0.25^{NS}
7	400	3,663	24	2.28 ± 0.51^{NS}	1.58 ± 0.43^{NS}
8	426	3,834	24	2.13 ± 0.46^{NS}	1.32 ± 0.31^{NS}
9	447	4,023	24	1.72 ± 0.42^{NS}	1.55 ± 0.40^{NS}
Mean				1.80 ± 0.15	0.92 ± 0.10

*Kolmogorov–Smirnov test for normal or nonnormal distribution of the variables. *P < 0.05: distribution significantly different from the normal distribution; NS: non-significant. [&]ATU: accumulated thermal units (days * °C above 0°C).*

Table 2. Skin spottiness density (number of dark spots/cm^2) on the back of juvenile rainbow trout at different developmental stages. Source: Colihueque (2010b)

PIT-tag individual	Developmental stages (days post-fertilisation)									Difference (final-initial)	Pearson correlation (r)
	282	303	318	330	365	380	400	426	447		
0675-1F76	2.67	3.11	0.00	5.78	8.44	7.56	7.11	7.56	3.56	0.89	0.51
068C-42FC	0.00	0.00	1.33	0.00	0.44	0.89	0.89	1.33	0.44	0.44	0.48
068C-526E	0.00	0.00	0.00	0.00	0.00	0.89	0.44	0.44	0.00	0.00	0.43
0698-22B6	4.89	3.11	8.00	8.89	4.00	5.78	4.89	5.33	4.00	-0.89	-0.22
0698-2D41	0.44	0.00	0.44	0.00	1.78	1.78	0.44	0.00	0.00	-0.44	0.01
0698-37DD	0.44	0.89	0.44	3.56	1.33	0.89	0.89	0.44	1.33	0.89	-0.04
0698-4199	0.00	0.44	2.22	1.78	4.00	3.11	2.22	3.56	2.22	2.22	0.67
069C-559F	0.00	0.00	1.33	1.33	0.00	0.44	0.00	0.44	0.00	0.00	-0.24
069C-5EE5	1.33	0.00	0.00	0.44	1.33	0.00	0.44	1.78	0.44	-0.89	0.18
069C-6826	0.00	1.33	2.22	4.44	3.56	6.67	8.89	5.78	8.89	8.89	0.90
069C-687B	0.00	0.00	2.22	0.00	0.00	0.00	0.00	0.44	0.00	0.00	-0.20
069C-6FAD	1.78	0.44	0.89	1.78	1.78	2.67	0.89	0.44	1.33	-0.44	-0.08
069C-7F8C	0.00	0.00	0.00	0.00	0.00	0.00	1.00	0.00	0.00	0.00	0.26
069C-7FAD	1.78	1.78	4.00	0.00	3.11	3.11	2.67	3.56	2.22	0.44	0.32
069C-81BA	0.00	1.33	0.44	4.44	4.89	3.56	2.67	3.11	1.78	1.78	0.39
069C-863F	0.44	0.00	1.78	4.89	3.11	1.78	3.56	1.78	4.00	3.56	0.51
069C-8B90	0.00	0.00	0.00	0.00	0.00	0.00	1.33	0.89	0.89	0.89	0.76
069C-D108	0.00	0.00	0.00	0.00	0.89	0.44	0.89	0.00	0.89	0.89	0.62
069C-D400	0.00	0.00	2.22	1.33	2.22	0.89	1.78	0.89	0.89	0.89	0.24
069C-D66D	0.00	0.44	0.00	0.89	0.44	0.89	0.00	0.44	0.89	0.89	0.40
069C-DOAC	0.89	1.33	3.56	0.00	1.78	3.56	1.78	1.78	1.33	0.44	0.13
069C-E34C	0.44	0.89	0.44	3.56	0.44	2.67	2.67	4.00	0.00	-0.44	0.28
069D-OEF2	2.22	3.11	1.33	1.33	1.33	3.11	1.33	0.89	1.78	-0.44	-0.35
06A9-E353	0.44	3.11	10.67	7.56	6.67	4.89	8.00	6.22	4.44	4.00	0.22

Table 3. Size of spots (percentage of dark spots/cm^2) on the back of juvenile rainbow trout at different developmental stages. Source: Colihueque (2010b)

PIT-tag	Developmental stages (days post-fertilisation)									Difference	Pearson
individual	282	303	318	330	365	380	400	426	447	(final-initial)	correlation (r)
0675-1F76	0.49	0.71	0.00	1.82	4.71	3.33	4.58	3.96	3.78	3.29	0.83
068C-42FC	0.00	0.00	0.22	0.00	0.13	0.44	0.31	0.67	0.18	0.18	0.67
068C-526E	0.00	0.00	0.00	0.00	0.00	0.31	0.22	0.44	0.00	0.00	0.55
0698-22B6	1.02	0.76	3.42	5.78	2.13	4.40	7.91	4.84	3.42	2.40	0.53
0698-2D41	0.04	0.00	0.13	0.00	0.36	0.84	0.13	0.00	0.00	-0.04	0.10
0698-37DD	0.09	0.36	0.22	1.47	0.49	0.58	0.93	0.13	2.62	2.53	0.53
0698-4199	0.00	0.09	0.76	0.58	1.87	1.42	1.56	2.00	1.42	1.42	0.85
069C-559F	0.00	0.00	0.27	0.22	0.00	0.18	0.00	0.13	0.00	0.00	-0.13
069C-5EE5	0.18	0.00	0.00	0.09	0.67	0.00	0.18	0.80	1.87	1.69	0.73
069C-6826	0.00	0.22	0.44	1.64	1.42	3.29	5.56	3.69	8.80	8.80	0.90
069C-687B	0.00	0.00	0.40	0.00	0.00	0.00	0.00	0.22	0.00	0.00	-0.04
069C-6FAD	0.27	0.09	0.13	0.71	0.31	1.16	0.36	0.22	0.80	0.53	0.40
069C-7F8C	0.00	0.00	0.00	0.00	0.00	0.00	0.50	0.00	0.00	0.00	0.26
069C-7FAD	0.31	0.53	2.18	0.00	2.31	2.09	1.96	3.47	2.13	1.82	0.74
069C-81BA	0.00	0.22	0.09	1.02	1.69	1.91	1.42	1.73	1.69	1.69	0.86
069C-863F	0.18	0.00	0.62	1.42	1.82	0.76	2.44	0.89	2.31	2.13	0.71
069C-8B90	0.00	0.00	0.00	0.00	0.00	0.00	0.40	0.58	0.62	0.62	0.85
069C-D108	0.00	0.00	0.00	0.00	0.71	0.13	0.49	0.00	0.71	0.71	0.59
069C-D400	0.00	0.00	0.62	0.49	1.02	0.22	1.11	0.36	0.53	0.53	0.43
069C-D66D	0.00	0.09	0.00	0.18	0.22	0.49	0.00	0.22	1.20	1.20	0.67
069C-DOAC	0.09	0.22	0.67	0.00	0.40	1.20	0.76	0.71	0.58	0.49	0.58
069C-E34C	0.04	0.09	0.04	1.16	0.09	1.42	1.02	2.22	0.00	-0.04	0.44
069D-OEF2	0.36	1.16	0.44	0.53	0.36	2.58	1.02	0.62	1.02	0.67	0.27
06A9-E353	0.09	0.53	2.89	2.04	2.93	1.64	5.07	3.69	3.51	3.42	0.77

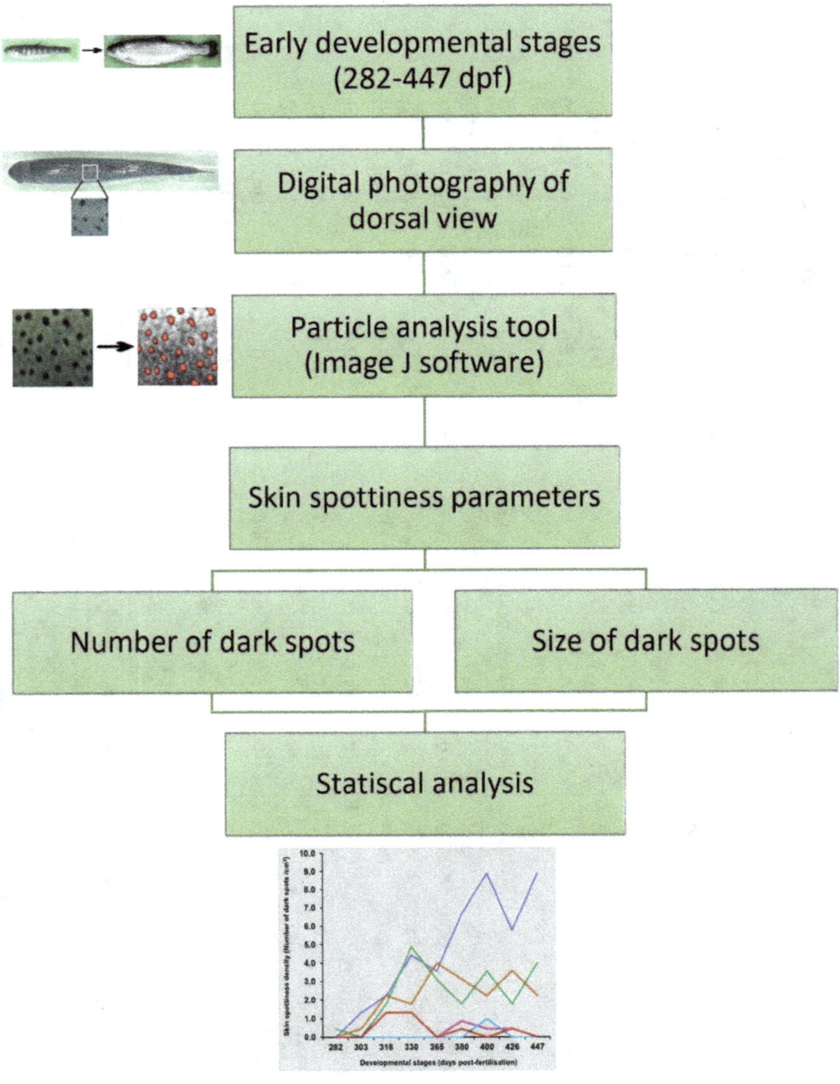

Fig. 1 The computer vision framework based on computer-based image analysis used to analyse skin spotting in juvenile rainbow trout.

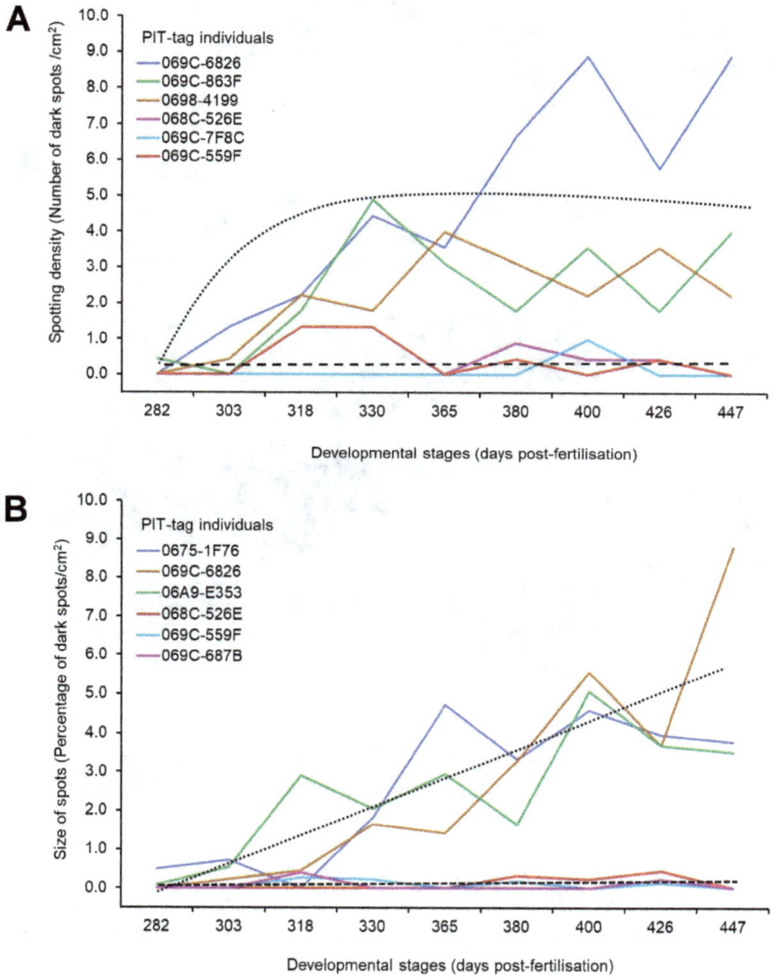

Fig. 2 Variations in dorsal skin spotting—(A) the number of dark spots and (B) the size of dark spots—in rainbow trout across different developmental stages. One group of individuals with increasing skin spottiness over time and another remaining entirely unspotted throughout development are shown. In both groups, the tendency lines are indicated in black. Source: Colihueque (2010b)

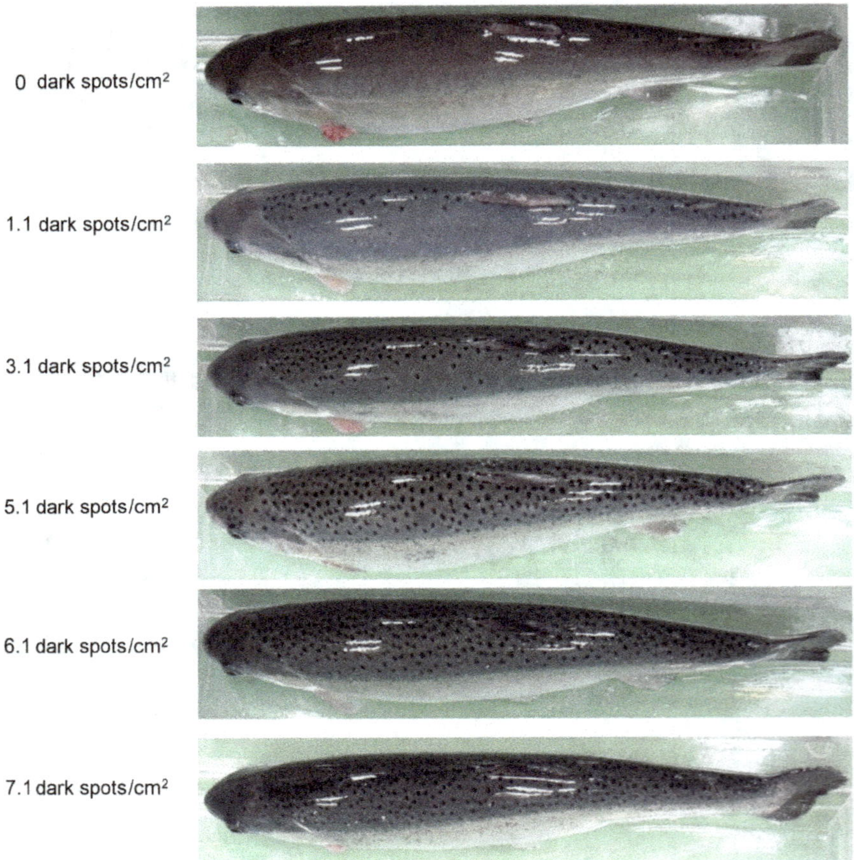

Fig. 3 Inter-individual variation in the skin spottiness on the back of rainbow trout at the juvenile stage. A full view of the back with different numbers of dark spots is shown. All fish are the same age (426 days post-fertilisation). Source: Colihueque (2010b)

Chapter 3.

Skin spotting variation associated with biometric and reproductive parameters in naturalised populations of rainbow trout, *Oncorhynchus mykiss*, from southern Chile

Structured abstract

Background: skin pigmentation in fish is involved in various biological processes. In salmonids, the interactions of skin spottiness with biometric and reproductive parameters are mostly unknown, especially in naturalised populations influenced by different ecological factors. **Objective**: to associate skin spottiness variation with biometric and reproductive parameters in broodstocks of naturalised rainbow trout populations from southern Chile. **Method**: the number of dark spots below the lateral line was assessed in male and female broodstocks of rainbow trout from two reproductive seasons, years 2007 and 2012, and then this dataset was associated with biometric and reproductive parameters of the same individuals, using multivariate data analysis in the form of stepwise forward multiple regression. **Results**: male body weight from year 2007 had a significant negative influence on number of dark spots below the lateral line ($P < 0.01$), while the reproductive variables egg diameter and spawning time of females from years 2007 and 2012, respectively, had a significant positive influence on this parameter ($P < 0.01$). **Conclusion**: our results indicate that there are male biometric parameters with a significant negative influence on skin spottiness. Our data also reveal that some reproductive parameters have a positive influence on skin spottiness. The identification of these reproductive parameters related to spottiness variation would reflect the reproductive quality of female broodstocks living in natural environments.

Keywords: rainbow trout, naturalised population, skin spottiness, biometry, reproduction, multiple regression

Note: Reproduced from Colihueque et al. 2018. The Open Ecology Journal 11, 25-37, DOI: 10.2174/1874213001811010025 (CC BY 4.0)

1. Introduction

Several studies have shown that skin pigmentation in vertebrates is a relevant character affecting different biological processes, such as the ornamentation of individuals, female mating preferences and the reproductive success of breeders in terms of offspring quality (Nordeide et al. 2013). In fishes, for example, its significance has been well documented in relation to the issue of sexual selection, as occurs in wild–type guppy males. In these fish, the male color pattern can determine the female mating preferences, since this character enhances the attractiveness of males during courtship (Kodric-Brown 1985, Houde and Endler 1990, Evans et al. 2004, Sathyan et al. 2013). In salmonids, a group that presents a characteristic species–specific pigmentation pattern (Colihueque 2010), studies have established that this character has a close relationship with developmental, reproductive and physiological processes. For example, some studies found that skin color intensity (*i.e.*, redness) and body color are associated with offspring quality (Janhunen et al. 2011, Ramstad et al. 2010), and that the silver skin color is related to the smoltification capacity of different strains (Nichols et al. 2008). In addition, it has been reported that, in some species of this group, other skin pigmentation components, particularly the spotting variation, could be related with fish size (Aparicio et al. 2005) and with the different environmental features of the rivers inhabited by these populations. In the latter case, reflecting the role of skin pigmentation as a signal status that could determine the spatial arrangement of individuals along the river bed (Penteriani et al. 2015). Further studies have indicated that spotting variation in salmonids can be involved in behavior-related roles, such as stress responsiveness, given that highly spotted individuals are generally more resistant to stress than less spotted fishes (Kittilsen et al. 2009). Taken together, these data provide strong evidence that skin pigmentation in salmonids plays an important role in different biological processes and may provide clues to facilitate the analysis of these events.

The spottiness is a characteristic attribute of the skin pigmentation of rainbow trout, *Oncorhynchus mykiss* (Walbaum, 1792). In adult individuals of this species this character is made up of conspicuous small dark spots, mainly distributed on the flank (above and below the lateral line), on the back, including the operculum and the dorsal and caudal fins (Ade 1989). The available evidence indicates that the spottiness is not a constant character in rainbow trout, given that it may vary within and between populations, *e.g.*, in number both below the lateral line (Islam et al. 1973, Tack 1973) and on the back (Colihueque et al. 2011, 2014), or in size on the flanks (Kause et al. 2003). Thus, it is possible to find trouts within the same stock with a wide variation in the number of dark spots. For example, on the back, which may result in individuals with scarce or no spotting (< 0.5 spots/cm^2), in contrast to others that can present a highly spotted skin (> 5.5 spots/cm^2) (Colihueque et al. 2011). To date, the origin of this variation in the rainbow trout is unclear, however, it has been reported that the number of dark spots may vary in function of fish size and spawning time for the spottiness below the lateral line (Islam et al. 1973), and with age for the spottiness on back (Colihueque et al. 2011). In this species additional evidences also support the possible effects of genetic factors on the variation of the number of dark spots, either below the lateral line (Islam et al. 1973, Tack 1973) or on the back (Díaz et al. 2011). Most studies known to date on the analysis of the variation of the number of dark spots in rainbow trout have been carried out in cultured populations (Islam et al. 1973, Tack 1973, Colihueque et al. 2011, 2014), being their occurrence in natural or naturalised population of this species largely unknown. In natural populations of salmonid fishes, the evidence indicates a variation of this character is associated with biometric and environmental factors. For example, in brown trout the number of dark spots distributed below the lateral line is positively associated with fish size (Aparicio et al. 2005) and in graylings, the spotting level on flank presents a clear interaction with different environmental features of the river that they inhabit (Penteriani et al. 2015).

The rainbow trout was introduced to our country in 1905 (Golusda 1927). Currently, several naturalised populations of this species are found from northern to southern Chile inhabiting different river and lakes. In this body of waters, the rainbow trout has importance for recreational fishing purposes. The naturalised populations of rainbow trout inhabiting lakes from southern Chile, called Araucanian lakes, are adapted to particular limnic conditions given that these body waters present special environmental characteristics, such as a) glacial origin with volcanic influence in the soils, b) high deepness, c) thermal stratification, d) high in water transparency, and e) oligotrophic with very low salt/nutrient concentrations (Campos 1984, Arismendi et al. 2011). In addition, within these lakes the naturalised populations of rainbow trout coexist with several native fishes and other introduced salmonids, most of them occupying the same habitat (Soto et al. 2006). Although some studies on the biology of naturalised populations of Araucanian lakes have been reported, such as those referring to the characterization of reproductive parameters and reproductive cycle (Wetzlar 1979, Soto et al. 2002), there have been no previous studies on their skin pigmentation. These populations provide an ideal opportunity to examine the relationship, for example, between the spottiness and reproductive parameters in nature to identify the factors that may contribute to spotting deviation. The spotting variation could be due to several factors, some of them being the heterozygosis of genes that control the trait, the interaction of individuals with different environmental features, the existence of different ecological pressures, sexual selection, and their correlation with behavioural and physiological traits, or a combination of the above. To assess whether the spotting level on the skin is associated with biometric and reproductive parameters of rainbow trout populations in nature, we investigated the relationship between the number of dark spots below the lateral line with different biometric and reproductive parameters in breeders of this species collected from a naturalised population.

This paper presents the result of the relationship assessed by stepwise multiple regression analyses between the number of dark spots below the lateral line with biometric and reproductive parameters by analysing male and female breeders from a naturalised population of rainbow trout that inhabit a lake from southern Chile. We analysed breeders from two reproductive seasons to obtain a more complete view of the possible trends that could emerge among these parameters in nature.

2. Material and methods

2.1 Fish collection and spawning

Sexually mature specimens were collected from different inlet streams of Calafquén Lake (39°31'S; 72°10'W, Villarrica district, La Araucanía Region, Chile), when the trouts carried out their reproductive run occurred from October to November, during the spawning season of the years 2007 and 2012. Next, they were reared at Chiguaico fish hatchery (Lican Ray City, Villarrica district) at a temperature that ranged from 7 to 8 °C with a constant flow of water at 10 L/s and a culture density of 15 kg/m^3 in a raceway–type pond. The collected female and male specimens were measured for body weight (BW) and total length (TL) to the nearest 5 g and 0.1 cm, respectively. The condition factor (CF) according to the following formula was also calculated:

$$CF = (BW/TL^3) \cdot 100$$

Means and ranges of biometric parameters BW, TL and CF of collected specimens are presented in Table 1. Ripe females were spawned individually and before handling they were anesthetised with a solution of benzocaine at 100 mg/L. Next, eggs were fertilised by a pool of diluted milt in Billard's activator (Billard 1992) from four to six males. Fertilised eggs from individual females were incubated separately in plastic baskets within a shallow-trough incubator. Data was recorded for the following parameters: 1) Fertilisation rate (FR) (%), 2) egg diameter (ED) (mm), 3) total fecundity (TF) (eggs/female), 4) relative fecundity (RF) (eggs/kg female), 5) eyed egg survival (ES) (%), and 6) spawning time (SPT) (number of days elapsed from the first day of the year until the spawning date of female). Means and ranges of the reproductive parameters FR, ED, TF, RF, ES and SPT of collecting females are shown in Table 2. Females and males were also photographed with a digital camera to register the number of dark spots on the skin. Once breeders were used in the reproductive process, all were returned live to the Calafquén Lake. Percentage data of FR and ES variables were arcsine-transformed prior to the analysis to better conform the assumption of normality. Rainbow trout specimens analysed in this study were captured during a restocking program of Calafquén Lake carried out from 2006 to 2012, since this lake had suffered a major decline in its population of rainbow trout, due to overfishing and poaching. To comply with the regulations of the Chilean fishing law, repopulation is performed from fingerlings produced by the breeding stock of the same basin. These broodstocks are captured during their spawning migration between the lake and the tributary streams. Calafquén Lake is representative of the rainbow trout population of the Araucanian lakes, since the introduction of rainbow trout was carried out in all of them from the same state hatcheries in the past.

2.2 Spotting analysis

To perform the analysis of spottiness, the number of dark spots below the lateral line (NS-BLL) were counted in the area B of the body according to Qadri et al. (1959). This area corresponds to a rectangular zone of the skin located below the lateral line, in the posterior tip of the operculum and in the front base of the pelvic fin (Fig. 1). We selected this area given that it exhibits a high variation in number of dark spots on the skin in rainbow trout compared with others anteroposterior zones of this species according to previous studies (Islam et al. 1973). The NS-BLL were counted from digital photographs from the right side of the specimen. Only dark spots visible at naked eye were counted. Size and shape of the spots were not considered and fused spots were counted as one. The NS-BLL was log-transformed to achieve normality.

2.3 Statistical analysis

The normality of variables was determined by the Kolmogorov-Smirnov (K-S) test. Differences between mean of numbers of dark spots were calculated with a two-way analysis of variance (2–way ANOVA) followed by Tukey's multiple comparison tests to carry out post hoc pairwise comparisons of means (Sokal and Rohlf 1995). Forward selection stepwise multiple regression analyses were performed on NS-BLL with biometric (BW, TL and CF) and reproductive (FR, ED, TF, RF, ES and SPT) variables, as potential predictors to identify which variable, or combinations of variables, was most highly correlated with, and therefore possibly causally linked to, NS-BLL. Stepwise multiple regression is a useful tool for identifying combinations of independent variables that explain the most variation in another dependent variable (Sokal and Rohlf 1995). The criteria used for variable selection during forward stepwise regression were an F-statistic above 3.84 as entrance condition and an F-value of 2.71 as an exit criterion. The residuals of the final models were explored to verify the assumptions of normality, homogeneity and linearity. The absence of multicollinearity between independent variables was also verified through the variation inflation factor (VIF), based on the tolerance value of each variable, according to the formula VIF = 1/tolerance. When the value of the VIF was more than 5, the multicollinearity was considered serious and the variable was eliminated from the analysis (Ghani et al. 2010). In addition, to control the probability that a true null hypothesis had been incorrectly rejected as a consequence of multiple hypotheses testing, the entire data set was subjected to correlation analysis (Spearman's rank correlation) between response and independent variables. This enabled us to apply a bootstrap procedure for correction of P-values, a re-sampling method that is recommended for correlated outcomes (Westfall and Young 1993). For this analysis an $\alpha = 0.05$ was used as the significant threshold. We also calculated Cohen's d statistic to estimate the effect size, defined as the degree at which the null hypothesis is false (Cohen 1988). Cohen's d statistic indicates that the greater the effect size, the greater the degree to which the phenomenon studied is present, *i.e.*, this statistic estimates the magnitude of an effect of interest. Thus, d values between 0.5 and 0.8, and above 0.8 are considered to be medium and large effects, respectively. Statistical analyses were carried out using the STATISTICA program, version 5.1 (StatSoft Inc., Tulsa, USA, 1996) and the MATLAB R2010a program (The MathWorks, Inc., Natick, USA, 2010).

3. Results

The number of dark spots varied widely in both reproductive periods, from 5 to 134 in females and from 0 to 146 in males from the 2007 year, and from 4 to142 in females and from 5 to 230 in males from the 2012 year (Table 3). Examples of male and female breeders of the 2007 year showing the extreme spotting variation are shown in Fig. 2. Moreover, this parameter showed normal distribution in males and females of both reproductive periods (K-S test, $P > 0.10$) (Fig. 3). Females from the 2007 year, displayed a higher mean number of dark spots than males (39.7 ± 26.1 and 34.6 ± 34, respectively), meanwhile the opposite was observed in females and males in the year 2012 (39.4 ± 29.8 and 43.9 ± 47.9, respectively) (Table 3). However, among these means, there were no significant differences (2-way ANOVA, $F_{3/112} = 0.24$, $P > 0.05$) either among sex within the same year or among sex between different years. It was noteworthy that the number of dark spots showed more variation in males than females in both years (males CV = 0.98–1.09 vs. females CV = 0.63–0.82). Indeed, in the year 2012 the number of dark spots in females exhibited less range (from 4 to142) than in males (from 5 to 230).

Stepwise multiple regression for biometric parameters revealed a significant relationship, ranging from medium to large effects, in broodstocks during year 2007. Thus, results indicated a statistically significant relationship, and medium to large effects, between NS-BLL and BW for males ($P < 0.01$, $d = 1.71$) and NS-BLL and TL for females ($P < 0.05$, $d = 0.73$) (Table 4). BW in males presented a medium negative correlation with NS-BLL ($\beta = -0.6496$), while TL in females showed a weak positive association with NS-BLL ($\beta = 0.3434$). R^2 indicated that the model explained 42.1% and 11.7% of the variability in the NS-BLL of males and females, respectively (Table 5). Bootstrap correction of P-values of Spearman correlation showed that only the relationship between NS-BLL and BW for males ($r_s = -0.5166$, $P = 0.0337$, $P_{BOOTS} = 0.0371$) was significant at $P < 0.05$. Thus, BW in males had the greatest influence on the response variable NS-BLL. For males 2012 no significant correlations were registered between NS-BLL and any biometric parameters ($P > 0.05$). In the case of BW a negative correlation was found ($\beta = -0.2794$), but without significance (exact P-value = 0.1860). Stepwise multiple regression for reproductive parameters revealed a significant relationship for females during the years 2007 and 2012 (Table 4). This analysis indicated the selection of only one of the six independent variables tested in the model, which had a significant association, and large effects, with NS-BLL that corresponded to ED for females in year 2007 ($P < 0.01$, $d = 1.0$) and SPT for females in year 2012 ($P < 0.01$, $d = 0.96$). In both cases, the correlations were positives and medium between the response variable and the independent variable ($\beta = 0.4487$ and $\beta = 0.4309$, respectively). R^2 indicated that 20.1% and 18.5% of variations in the response variable NS-BLL occurred because of changes in the independent variables ED and SPT, respectively (Table 5). Bootstrap correction of P-values of Spearman correlation showed that the relationships between NS-BLL and ED for females year 2007 ($r_s = 0.5328$, $P = 0.0011$, $P_{BOOTS} = 0.0021$), and NS-BLL and SPT for females year 2012 ($r_s = 0.4166$, $P = 0.0067$, $P_{BOOTS} = 0.0113$) were significants at $P < 0.05$. This result indicates that ED and SPT are important factors in the response variable NS-BLL. The equations of the fitted models were

Females year 2007: $$\text{NS-BLL} = -0.9010 + 0.4461 \cdot \text{ED}$$

Females year 2012: $$\text{NS-BLL} = -3.8494 + 0.0186 \cdot \text{SPT}$$

4. Discussion

Skin pigmentation in fishes has been related to biological and environmental parameters to try to understand if the variation of this character, either within or among populations, has a biological significance. To date these studies have been able to clarify the participation of this phenotype in several processes, for example, in predation avoidance (Endler 1980), in female mating preferences (Kodric-Brown 1985, Houde and Endler 1990, Evans et al. 2004), in phenotypic plasticity that may facilitate invasions in nature (Westley et al. 2013), in cryptic camouflage (Donnelly and Dill 1984, Donnelly and Whoriskey 1993), in status signals related to the spatial arrangement of individuals along the river bed (Penteriani et al. 2015), and, also, in stress response of individuals (Kittilsen et al. 2009). In spite of this progress, few studies have been conducted on the role of skin pigmentation in relation to the reproductive success of broodstocks, although the scarce data available on this issue suggest its possible role in spawning displays and reproductive success (Wedekind et al. 2008, Janhunen et al. 2011). This study contributes to understanding the influence of skin pigmentation, particularly skin spottiness, on reproductive and biometric patterns observed in two naturalised populations of rainbow trout. Through this analysis we provide new data on the possible trends that could emerge among these parameters in nature, in which particular intrapopulation interactions may have taken place in response to different ecological factors.

In this study, we support that below the lateral line (area B) the number of dark spots is also a variable character in naturalised populations of this species from southern Chile. Thus, for example in the 2007 stock, this parameter varied from 4 to 142 and from 5 to 230 dark spots in females and males, respectively. As far as we know, to date no data on naturalised population from Chile and other countries has been published on this character to compare our results. However, taken into account the data available for the cultured population of this species, our results have a good concordance. Indeed, Islam et al. (1973) reported hatchery populations from Japan with a number of dark spots in the area B ranging either from 6 to 338, or from 12 to 316. In addition, our data also agree with reports from the other cultured population, as is the case of a German population where this character varied from 0 to 250 (Tack 1973). In summary, our data along with previous studies indicate that the number of dark spots below the lateral line of the rainbow trout is mostly a polymorphic character, a phenomena that are displayed by cultured and naturalised populations of this species.

The variation in the number of dark spots in the naturalised population of rainbow trout studied here is an interesting result that will require more in depth investigation to clarify their origin. However, this variation appears to be not sex-linked given that males and females analysed in this work from both reproductive periods did not present significant differences in this parameter. Other factors, mostly studied in cultured populations of this species, although not analysed in this work, may involve the effect of genetic, environmental and even, behavioural factors. For instance, the strain-specific variation in the number of dark spots below the lateral line (Islam et al. 1973), and the transmission of this character to the offspring in experimental crosses of progenitors with contrasting number of dark spots (Tack 1973), represent findings that support the effect of genetic factors in rainbow trout. In line with these evidences, more recent data obtained in this species also indicate the effect of a polygenetic determination for the dark spotting on the skin, whose heritability (h^2) could reach a value ranging from $h^2 = 0.16$ (Díaz et al. 2011) to $h^2 = 0.45$ (Kause et al. 2003). In addition, the action of environmental factors on the spotting level in rainbow trout would also occur based on the evidence documented in other salmonids. Particularly, in the brown trout, where it has been demonstrated that the variance of the number of dark spots below the lateral line could be subjected to a significant environmental effect (Blanc et al. 1982), and in graylings, whose spotting level in the flank is positively associated with particular environmental features of the river, such as water velocity, water depth and water transparency (Penteriani et al. 2015). Other studies even support the effect of behavioural factors, given that in rainbow trout and Atlantic salmon the variation in spots number is related to the stress response of individuals (Kittilsen et al. 2009), being this capacity, higher in spotted fish than in non-spotted fish.

In females of the year 2007, we found a weak positive association between the number of dark spots below the lateral line and total length. This pattern is discordant with previous studies since, in rainbow trout, a clear relationship among these variables has been described for specimens above a certain size (> 30 cm) (Islam et al. 1973). Our finding also disagrees with reports from other salmonid fishes, such as the brown trout, since total length in this species is significantly associated with the number of dark spots located below the lateral line (Aparicio et al. 2005), with those distributed in the ventral zone (Blanc et al. 1982), and with the total number of red spots on the body (Kocabas et al. 2011). In contrast, the males analysed in this study, especially those of the year 2007, exhibited a strong inverse relationship between both parameters, *i.e.*, larger specimens tend to have a lower number of dark spots below the lateral line than small ones. It is possible that this association may be related to the different development stages or ages of males analysed if we consider that the spotting level decrease across the development. However, the available evidence in the rainbow trout indicates that on the contrary, this interpretation would not be consistent given that the spotting level of this species increases steadily across the development, such as have been observed at juvenile stage (Colihueque 2011). Therefore, it is likely that other factors could be involved in the origin of this pattern, being environmental factors an interesting hypothesis to be considered in the light of the evidence obtained in other salmonids. For example, in natural populations of graylings, Penteriani et al. (2015) have supported the interaction of the spotting level on flank with different environmental features of the river where they inhabit, such as water velocity, water depth and water transparency. Indeed, individuals with high spottiness level exhibit a positive and significant association to fast waters or with greater depths. Interestingly, these interactions also reveal that the spotting level may constitute a status signal related to the spatial arrangement of individuals along the river bed. Although further analyses on this issue are required to clarify the possible effects of environmental factors on the spotting level in naturalised populations of rainbow trout, in our case it should be noted that specimen analysed were collected at spawning season from the small inlet streams of Calafquén Lake during their migration period. As it has been observed in this species (Knapp and Vredenburg 1996), the spawning behaviour is a complex process that involves the interaction of spawning females with particular characteristics of the river to construct their redds, such as water velocity, water depth, and substrate size. Moreover, in this process the interaction of females with males is also important for fertilisation success, especially, if we keep in mind that in natural conditions ripe males are 3-9 times more abundant than ripe females (Knapp and Vredenburg 1996). Thus, it is possible that the spotting variation of males that is negatively associated with size may constitute some class of signal to address the interaction by spawning females.

The relationship between skin color and reproductive parameters in rainbow trout has received little attention. In vertebrates, including fishes, this approximation has been analysed in the context of the effect of female ornamentation on offspring quality, and therefore, as a factor affecting fitness (reviewed by Nordeide et al. 2013). These studies indicate that different vertebrate ornamentation attributes (*e.g.*, plumage coloration, size of black feather spots, skin color and body color) are related positively or negatively at different levels; for example, with the survival and body mass of the offspring and with the size, mass and color of eggs. In particular, the effects of spotting level on fitness were well documented in *Rana pipiens* several years ago (Merrel 1972). It has also been observed that a particular female ornamentation may affect the female's own reproductive parameters, such as spawning time and fecundity (Roulin et al. 2001, Janhunen et al. 2011). In our case, results indicating that the existence of some reproductive parameters in female rainbow trout broodstocks, particularly egg diameter and spawning time, were significantly associated with the number of dark spots below the lateral line, suggest that this skin color attribute may be linked with female fitness. Thus, the spotting level could be interpreted as an indicator that reflects the reproductive quality of females, especially if we take into account that egg diameter (Brooks et al. 1997) and spawning time (Siitonnen and Gall 1989, Quinn et al. 2002) are important traits for salmonid reproduction. In addition, since our study analysed a naturalised population of rainbow trout, it is possible that ecological factors may be involved in the origin of this relationship. This possibility is not unlikely, in view of ecological studies on different fishes which indicate that greater reproduction success in females under natural conditions appears to depend on egg size, since larger eggs provide survival and growth advantages to the offspring during their first few days of life (Brooks et al. 1997). In addition, our findings on the association between spottiness and spawning time, agree with other previous studies that reveal the existence of this relationship in cultured populations of rainbow trout (Islam et al. 1973). They would also concur with data available on other vertebrates, such as the barn owl, where this association occurs in relation to the size of dark spots (Roulin et al. 2001). However, in contrast to Islam et al. (1973), we observed a positive relationship between the numbers of dark spots below the lateral line and spawning time, *i.e.*, females with higher numbers of dark spots tend to spawn later in the reproductive season than females with lower numbers of spots. Although, based on the present data, it is difficult to clarify the origin of this discordance between both data sets, it is possible that strain differences, management conditions, selection or genetic factors may explain, in part, these results. This is not unlikely, given that Islam et al. (1973) studied cultured strains in Japan, which that are usually subject to intense management in fish farms and, in contrast to a free-living population, are reared under controlled environmental conditions. These conditions are very different to those experienced by the naturalised populations of rainbow trout studied in this work, given that they inhabit a lake in southern Chile (Calafquén Lake), with particular limnic conditions related to special water characteristics (Campos 1984, Arismendi et al. 2011). In the case of selection and genetic factors, their effect on spawning time and spottiness would be expected to produce considerable variation among strains. This is possible given that, in rainbow trout, spawning time (Siitonnen and Gall 1989, Su et al. 1999) and spotting level (Kause et al. 2003) characters have a significant genetic determination, and therefore, may respond to selection pressure, which would alter their distribution in populations. In fact, Kause et al. (2003) have demonstrated that through artificial selection of the spottiness level of a cultured rainbow trout strain, it is possible to respond rapidly to selection pressure after a few generations.

Further studies on spotting variation in other naturalised populations of rainbow trout from Chile will be important to support the relationship of this parameter with biometric and reproductive parameters reported in this work. The analysis of the spotting variation in these populations, for example, according to the habitat characteristics, it will also be useful to add more data to clarify this issue.

5. Conclusion

No significant differences were observed in the mean number of dark spots below the lateral line between males and females in both reproductive seasons. This result suggests that spottiness below the lateral line is not sex-specific. Stepwise multiple regression analyses revealed that the body weight of males from year 2007 has a significant negative influence on the number of dark spots below the lateral line, while the reproductive variables egg diameter and spawning time of females from years 2007 and 2012, respectively, also have a relevant positive influence on this variable response. Overall, these results suggest that some biometric and reproductive variables may influence the skin spot variation of rainbow trout broodstocks living in natural environments.

6. Acknowledgements

The suggestions and comments of all those who helped to improve the final version of this manuscript, are gratefully acknowledged. We are also indebted to Chiguaico fish farm for the hatchery facilities and Lican Ray Fish and Game Club for its assistance in the field work during collection of fish. This study was partially financed by Subsecretaría de Pesca y Acuicultura, Ministerio de Economía, Fomento y Turismo, Gobierno de Chile, Project 2012-34-FAP-5. This work is dedicated to the memory of Alejandro Koffmann O'Reilly, an outstanding promoter of the rainbow trout restocking program of Lake Caláfquen.

7. References

Ade, R (1989) *Trout and salmon handbook*. Facts on File, Inc., New York.

Aparicio, E, García-Berthou, E, Araguas, RM, Martínez, P, & García-Marín, JL (2005) Body pigmentation pattern to assess introgression by hatchery stocks in native *Salmo trutta* from Mediterranean streams. *Journal of Fish Biology*, 67, 931–49.[http://dx.doi.org/doi:10.1111/j.0022-1112.2005.00794.x]

Arismendi, I, Sanzana, J, & Soto, D (2011) Seasonal age distributions and maturity stage in a naturalized rainbow trout (*Oncorhynchus mykiss* Walbaum) population in southern Chile reveal an ad-fluvial life history. *Annales de Limnologie - International Journal of Limnology*, 47, 133-40. [http://dx.doi.org/doi:10.1051/limn/2011012]

Billard, R (1992) Reproduction in rainbow trout: sex differentiation, dynamics of gametogenesis, biology and preservation of gametes. *Aquaculture*, 100, 263–98.[http://dx.doi.org/doi:10.1016/0044-8486(92)90385-X]

Blanc, JM, Poisson, H, & Vibert, R (1982) Variabilité génétique de la ponctuation noire sur la truitelle Fario (*Salmo trutta* L.). *Annales de Génétique et de Sélection Animal*, 14, 225–36.

Brooks, S, Tyler, CR, & Sumpter, JP (1997) Egg quality in fish: what makes a good egg? *Reviews in Fish Biology and Fisheries*, 7, 387–416. [http://dx.doi.org/doi:10.1023/A:1018400130692]

Campos, H (1984) Limnological study of Araucanian lakes (Chile). *Verhandlungen Internationale Vereinigung Limnologie*, 22, 1319–27.

Cohen, J (1988) *Statistical power analysis for the behavioral sciences*. Lawrence Erlbaum Associates, New York.

Colihueque, N (2010) Genetics of salmonid skin pigmentation: clues and prospects for improving the external appearance of farmed salmonids. *Reviews in Fish Biology and Fisheries*, 20, 71–86. [http://dx.doi.org/doi:10.1007/s11160-009-9121-6]

Colihueque, N, Parraguez, M, Estay, FJ, & Diaz, NF (2011) Skin color characterization in rainbow trout by use of computer-based image analysis. *North American Journal of Aquaculture*, 73, 249–58. [http://dx.doi.org/doi:10.1080/15222055.2011.581578]

Colihueque, N (2014) Analysis of the coloration and spottiness of Blue Back rainbow trout at a juvenile stage. *Journal of Applied Animal Research*, 42, 474–80.[http://dx.doi.org/doi:10.1080/09712119.2014.883317]

Diaz, NF, Estay, F, Araneda, C, & Colihueque, N (2011) *Heritability of Blue Back skin colour and correlations with growth in rainbow trout, Oncorhynchus mykiss*. Proceedings of the Australian Society for Fish Biology 2011 Conference, Australian Society for Fish Biology, Townsville, Australia.

Donnelly, WA, & Dill, LM (1984). Evidence for crypsis in coho salmon, *Oncorhynchus kisutch* (Walbaum), parr: substrate colour preference and achromatic reflectance. *Journal of Fish Biology*, 25, 183–95. [http://dx.doi.org/doi:10.1111/j.1095-8649.1984.tb04865.x]

Donnelly, W, & Whoriskey, F (1993). Transplantation of Atlantic salmon (*Salmo salar*) and crypsis breakdow. *Canadian Special Publication of Fisheries and Aquatic Sciences*, 118, 25–34.

Endler, JA (1980). Natural selection on color patterns in *Poecilia reticulata*. *Evolution*, 34, 76–91. [http://dx.doi.org/doi:10.1111/j.1558-5646.1980.tb04790.x]

Evans, JP, Bisazza, A, & Pilastro, A (2004) Female mating preferences for colourful males in a population of guppies subject to high predation. *Journal of Fish Biology*, 65, 1154–59. [http://dx.doi.org/doi:10.1111/j.0022-1112.2004.00502.x]

Ghani, IMM, & Ahmad, S (2010) Stepwise multiple regression method to forecast fish landing. *Procedia - Social and Behavioral Sciences*, 8, 549–54. [http://dx.doi.org/doi:http://dx.doi.org/10.1016/j.sbspro.2010.12.076]

Golusda, P (1927) Aclimatación y cultivo de especies salmonídeas en Chile. *Boletín de la Sociedad de Biología de Concepción-Chile*, 1, 80–100.

Houde, AE, & Endler, JA (1990) Correlated evolution of female mating preferences and male color patterns in the guppy *Poecilia reticulata*. *Science*, 248, 1405–08. [http://dx.doi.org/doi:10.1126/science.248.4961.1405]

Islam, MA, Nose, Y, & Yasuda, F (1973) Number and distribution pattern of the black spots on body surface of rainbow trout. *Bulletin of the Japanese Society of Scientific Fisheries*, 39, 727–39.

Janhunen, M, Peuhkuri, N, Primmer, CR, Kolari, I, & Piironen, J (2010) Does breeding ornamentation signal genetic quality in Arctic charr, *Salvelinus alpinus*? *Evolutionary Biology*, 38, 68–78. [http://dx.doi.org/doi:10.1007/s11692-010-9100-9]

Kause, A, Ritola, O, Paananen, T, Eskelinen, U, & Mäntysaari, E (2003) Big and beautiful? Quantitative genetic parameters for appearance of large rainbow trout. *Journal of Fish Biology*, 62, 610–22. [http://dx.doi.org/doi:10.1046/j.1095-8649.2003.00051.x]

Kittilsen, S, Schjolden, J, Beitnes-Johansen, I et al. (2009). Melanin-based skin spots reflect stress responsiveness in salmonid fish. *Hormones and Behavior*, 56, 292–98.[http://dx.doi.org/doi:http://dx.doi.org/10.1016/j.yhbeh.2009.06.006]

Knapp, RA, & Vredenburg, VT (1996) Spawning by California Golden trout: characteristics of spawning fish, seasonal and daily timing, redd characteristics, and microhabitat preferences. *Transactions of the American Fisheries Society*, 125,519–31.[http://dx.doi.org/doi:10.1577/1548-8659(1996)125<0519:SBCGTC>2.3.CO;2]

Kodric-Brown, A (1985) Female preference and sexual selection for male coloration in the guppy (*Poecilia reticulata*). *Behavioral Ecology and Sociobiology*, 17, 199–205.[http://dx.doi.org/doi:10.1007/BF00300137]

Kocabas, M, Kayim, M, Can, E, Ates, M, Kutluyer, F, & Aksu, Ö (2011) Spotting pattern features in the brown trout (*Salmo trutta macrostigma*, T., 1954) population. *Scientific Research and Essays*, 6, 5021–24. [http://dx.doi.org/doi:10.5897/SRE11.1098]

Merrell, DJ (1972) Laboratory studies bearing on pigment pattern polymorphisms in wild populations of *Rana pipiens*. *Genetics*, 70, 141–61.

Nichols, KM, Edo, AF, Wheeler, PA, & Thorgaard, GH (2008) The genetic basis of smoltification-related traits in *Oncorhynchus mykiss*. *Genetics*, 179, 1559–75.[http://dx.doi.org/doi:10.1534/genetics.107.084251]

Nordeide, JT, Kekäläinen, J, Janhunen, M, & Kortet, R (2013) Female ornaments revisited - are they correlated with offspring quality? *The Journal of Animal Ecology*, 82, 26–38.[http://dx.doi.org/doi:10.1111/1365-2656.12021]

Penteriani, V, Delgado, MM, Clavero, M, & Ivanc, M (2015) Spottiness of European graylings (Actinopterygii, Salmonidae) correlates with environmental features and could signal the status of individuals. *Italian Journal of Zoology*, 82, 202–11. [http://dx.doi.org/doi:10.1080/11250003.2015.1018852]

Qadri, SU (1959) Some morphological differences between the subspecies of cutthroat trout, *Salmo clarki clarki* and *Salmo clarki lewisi*, in British Columbia. *Journal of the Fisheries Research Board of Canada*, 16, 903–22. [http://dx.doi.org/doi:10.1139/f59-064]

Quinn, TP, Peterson, JA, Gallucci, VF, Hershberger, WK, & Brannon, EL (2002) Artificial selection and environmental change: Countervailing factors affecting the timing of spawning by Coho and Chinook salmon. *Transactions of the American Fisheries Society*, 131, 591–98.[http://dx.doi.org/doi:10.1577/1548-8659(2002)131<0591:ASAECC>2.0.CO;2]

Ramstad, KM, Woody, CA, & Allendorf, FW (2010) Recent local adaptation of sockeye salmon to glacial spawning habitats. *Evolutionary Ecology*, 24, 391–411.[http://dx.doi.org/doi:10.1007/s10682-009-9313-5]

Roulin, A, Dijkstra, C, Riols, C, & Ducrest, A (2001) Female- and male-specific signals of quality in the barn owl. *Journal of Evolutionary Biology*, 14, 255–66. [http://dx.doi.org/doi:10.1046/j.1420-9101.2001.00274.x]

Sathyan, R, & Couldridge, VCK (2012) Female preference for blue in Japan blue guppies (*Poecilia reticulata*). *Environmental Biology of Fishes*, 96, 953–59. [http://dx.doi.org/doi:10.1007/s10641-012-0091-5]

Siitonen, L, & Gall, GAE (1989) Response to selection for early spawn date in rainbow trout, *Salmo gairdneri*. *Aquaculture*, 78, 153–61. [http://dx.doi.org/doi:http://dx.doi.org/10.1016/0044-8486(89)90029-X]

Sokal, RR, & Rohlf, FJ (1995) *Biometry: the principles and practice of statistics in biological research*. WH Freeman and Company, New York.

Soto, D (2002) *Estudio del ciclo reproductivo de las principales especies objetivo de la pesca deportiva en la X Región. Informe Final Proyecto FIP 2000-24.* Subsecretaría de Pesca y Acuicultura, Ministerio de Economía, Fomento y Turismo, Gobierno de Chile, Valparaiso, Chile.

Soto, D, Arismendi, I, Gonzalez, J et al. (2006) Southern Chile, trout and salmon country: invasion patterns and threats for native species. *Revista Chilena de Historia Natural*, 79, 97–117. [http://dx.doi.org/doi:10.4067/S0716-078X2006000100009]

Su, G-S, Liljedahl, L-E, & Gall, GAE (1999) Estimates of phenotypic and genetic parameters for within-season date and age at spawning of female rainbow trout. *Aquaculture*, 171, 209–20. [http://dx.doi.org/10.1016/S0044-8486(98)00494-3]

Tack, E (1973) Züchtungsfragen in der Forellenzucht. *Österreichs Fischerei*, 26, 77-84.

Wedekind, C, Jacob, A, Evanno, G, Nusslé, S, & Müller, R (2008). Viability of brown trout embryos positively linked to melanin-based but negatively to carotenoid-based colours of their fathers. *Proceedings of the Royal Society B: Biological Sciences*, 275, 1737–44.[http://dx.doi.org/doi:10.1098/rspb.2008.0072]

Westfall, PH, & Young, SS (1993) *Resampling-based multiple testing: Examples and methods for p-value adjustment*. John Wiley and Sons, New York.

Westley, PAH, Stanley, R, & Fleming, IA (2013). Experimental tests for heritable morphological color plasticity in non-native brown trout (*Salmo trutta*) populations. *PloS one*, 8, e8040. [http://dx.doi.org/doi:10.1371/journal.pone.0080401]

Wetzlar, HJ (1979) *Beitrage sur biologie und bewirtschaf tung von forellen (Salmo gairdneri und S. trutta) in Chile*. PhD Thesis, Albert-Ludwig Universitat, Contanza, Germany.

Table 1. Mean (± 1 SD), sample size (N) and range (in parenthesis) of biometric parameters registered for male and female breeders of naturalised rainbow trout population from southern Chile.

Parameter	Year 2007		Year 2012	
	Females (N = 34)	Males (N = 17)	Females (N = 41)	Males (N = 24)
Body weight (g)	1794.7 ± 608.7 (845–3150)	1074.1 ± 537.4 (495–2370)	2076.7 ± 604.5 (658–3644)	1532.6 ± 871.0 (424–4088)
Total length (cm)	51.5 ± 5.9 (40.5–63.1)	45.0 ± 7.6 (36.9–64.5)	54.5 ± 5.8 (37–65.5)	49.4 ± 8.4 (33.5–67.0)
Condition factor (CF)	1.3 ± 0.2 (1–2.1)	1.1 ± 0.1 (0.9–1.3)	1.3 ± 0.1 (0.7–1.5)	1.2 ± 0.1 (1.0–1.5)

Table 2. Mean (± 1 SD), sample size (N) and range (in parenthesis) of reproductive parameters registered for female breeders of naturalised rainbow trout population from southern Chile.

Parameter	Year 2007 (N = 34)	Year 2012 (N = 41)
Fertilisation rate (%)	93.3 ± 11.2 (50–100)	89.5 ± 9.2 (70–100)
Egg diameter (mm)	5.4 ± 0.3 (4.7–6.1)	5.3 ± 0.3 (4.5–6.1)
Total fecundity (eggs/female)	2934 ± 889.9 (1102–4930)	3456 ± 1298.1 (1332–7096)
Relative fecundity (eggs/kg female)	1735.6 ± 539.4 (918.7–2930.9)	1726.0 ± 568.9 (568.9–638.0)
Eyed egg survival (%)	84.3 ± 16.3 (36.8–97.8)	87.6 ± 10.9 (60.5–99.4)
Spawning time (days)	300.2 ± 14.5 (278–330)	286.5 ± 7.6 (274–302)

Table 3. Number of dark spots below the lateral line of male and female breeders of naturalised rainbow trout population from southern Chile. Comparison of means within and between sex was not significantly different according to 2-way ANOVA ($F_{3/112} = 0.24$, $P > 0.05$). N = sample size.

	Year 2007		Year 2012	
	Females (N = 34)	Males (N = 17)	Females (N = 41)	Males (N = 24)
Mean ± SD	39.7 ± 26.1	34.6 ± 34.1	39.4 ± 29.8	43.9 ± 47.9
Range	5-134	0-146	4-142	5-230
Coefficient of variation (CV)	0.66	0.98	0.76	1.09

Table 4. Final stepwise multiple regression models. Models show the effect of the biometric and reproductive variables on the number of dark spots below the lateral line in male and female broodstocks of a naturalised population of rainbow trout from southern Chile. SE = standard error of the estimate, β = regression coefficient, P = p-value with exact p-value indicated in parenthesis.

Parameters/Groups	Model	Coefficients	SE	β	t-value	*P*
Biometric parameters						
Males year 2007	Constant	1.9875	0.2075		9.5779	< 0.001 (0.0000)
	Body weight	-0.0006	0.0002	-0.6496	-3.3088	< 0.01 (0.0048)
Females year 2007	Constant	0.6014	0.4458		1.3492	> 0.05 (0.1868)
	Total length	0.0178	0.0086	0.3434	2.0681	< 0.05 (0.0468)
Reproductive parameters						
Females year 2007	Constant	-0.9010	0.8529		-1.0564	> 0.05 (0.2987)
	Egg diameter	0.4461	0.1571	0.4487	2.8400	< 0.01 (0.0078)
Females year 2012	Constant	-3.8494	1.7923		-2.1478	< 0.05 (0.0380)
	Spawning time	0.0186	0.0063	0.4309	2.9821	< 0.01 (0.0049)

Table 5. Model summary of the stepwise multiple regression. R = coefficient of multiple correlation, R^2 = coefficient of multiple determination, SE = standard error of the estimate, P = p-value with exact p-value indicated in parenthesis.

Parameters/ Groups	*R*	R^2	Adjusted R^2	SE	*F*	*P*
Biometric parameters						
Males year 2007	0.6496	0.4219	0.3834	0.3736	$F_{1/15} = 10.948$	< 0.01 (0.0048)
Females year 2007	0.3434	0.1179	0.0903	0.2938	$F_{1/32} = 4.2772$	< 0.05 (0.0468)
Reproductive parameters						
Females year 2007	0.4487	0.2013	0.1764	0.2796	$F_{1/32} = 8.0656$	< 0.01 (0.0078)
Females year 2012	0.4309	0.1857	0.1648	0.3005	$F_{1/39} = 8.8930$	< 0.01 (0.0049)

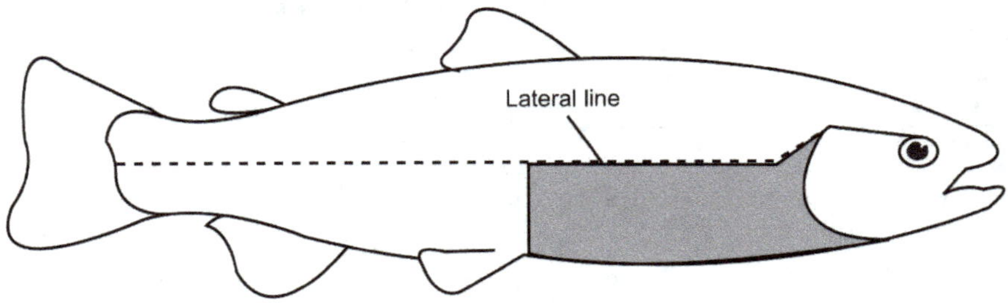

***Fig. 1** The body area below the lateral line (highlighted in gray) selected for the analysis of the number of dark spots in rainbow trout.*

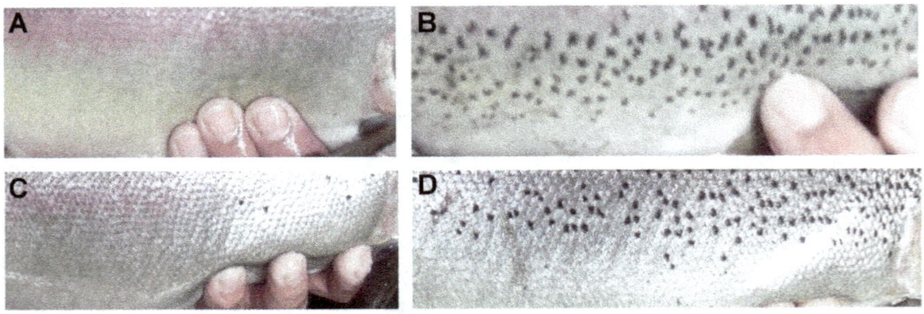

***Fig. 2** Images showing the wide spotting variation below the lateral line in breeders of a naturalised rainbow trout population from year 2007. A male with low spotted skin (< 20 dark spots), B male with high spotted skin (> 120 dark spots), C female with low spotted skin (< 20 dark spots), D female with high spotted skin (> 120 dark spots)*

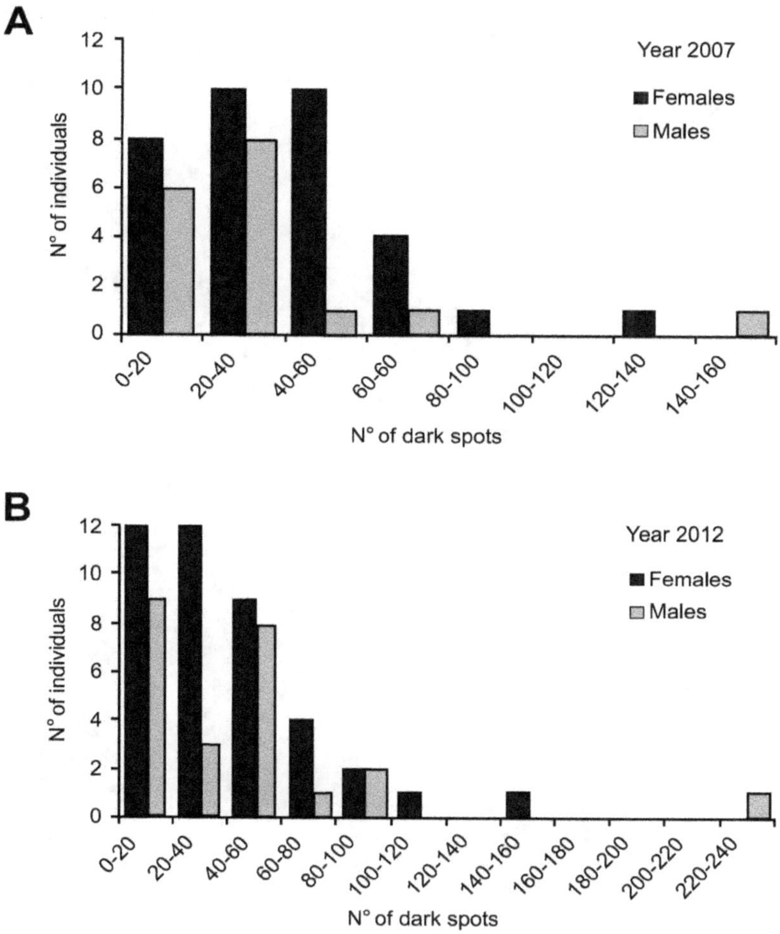

Fig. 3 Distribution of the number of dark spots below the lateral line in breeders of a naturalised rainbow trout population from southern Chile. A Males and females breeders of the spawning season from year the 2007, B males and females breeders of the spawning season from the year 2012.

Chapter 4.

Association between the blue color intensity of the skin with growth performance and skin spottiness in a blue-colored strain of rainbow trout (*Oncorhynchus mykiss*)

Structured abstract

Background: the blue-colored phenotypes have potential to strengthen the external appearance of rainbow trout for marketing due to their attractive skin coloration. However, further analyses are necessary to assess their overall performance for fish farming. **Objective**: to investigate a blue-colored strain of rainbow trout to determine whether growth performance and spotting variation of the skin, either in number or size, are related to the blue color intensity on the back skin (BCI). **Method**: a correlation analysis and comparisons among trout sorted in Low- and High-BCI categories were carried out in three blue-colored groups of rainbow trout. **Results**: in all groups, BCI was significantly and positively related to body weight ($0.01 < P < 0.05$). There was a significant and negative association between BCI and the number of dark spots in one group ($P < 0.001$), whereas no significant association was observed among BCI and dark spots size in any group ($P > 0.05$). A significant higher body weight in High-BCI (BCI ≥ 50%) than Low-BCI (BCI < 50%) category was observed in one group ($P < 0.05$). The same pattern was observed in two groups ($P < 0.05$) when lowest (1st) and highest (4th) quartiles of BCI were compared. Number of dark spots was significantly lower in High-BCI (≥ 50% and 4th quartile) than Low-BCI (< 50% and 1st quartile) categories ($P < 0.01$ and $P < 0.05$, respectively) in one group. Dark spots size presented no significant differences among BCI categories ($P > 0.05$) in any group. **Conclusion**: results suggest that trout with more intense blue color on skin have better growth performance than less blue-colored ones, and with a relatively less spotted skin. These characteristics may represent an interesting advantage of using this class of trout in intensive fish farming.

Keywords: aquaculture, rainbow trout, blue-colored strain, growth performance, skin spottiness

Note: Reproduced from Colihueque and Estay (2018). The Open Agriculture Journal 12, 251-261, DOI: 10.2174/1874331501819010251 (CC BY 4.0)

1. Introduction

The blue-colored morph of rainbow trout *Oncorhynchus mykiss* (Walbaum, 1792) is a distinctive phenotype among the various color morphs reported in this species, which is generally characterised by a blue back, a whitish belly, and a conspicuously silvery body. This color morph has been reported in various cultured populations around the world and includes the iridescent metallic blue variant (Kincaid 1975), the cobalt variant (Yamazaki 1974, Oguri 1983), the new blue variant (Blanc et al. 2006), and the Blue Back strain (Colihueque et al. 2011). In intensive farming, the blue-colored phenotype of rainbow trout is considered an attractive trait to satisfy consumer demands of certain markets. Thus, the use of this phenotype in production has great potential. This is the case of the iridescent metallic blue variant studied by (Kincaid 1975), that, in addition to its attractive external appearance, also presents a better growth performance than common trout, normal fertility, and the egg survival does not differ from that observed in control groups. As a result, this strain has been considered suitable for farming. However, the advantages reported for the iridescent metallic blue variant of rainbow trout do not appear to apply, generally, to other blue-colored morphs of this species, given that this condition may show detrimental pleiotropic effects on growth, survival and condition factor. In fact, survival and body weight are about 25% lower for the new blue variant than the wild-type at the juvenile stage (Blanc et al. 2006), and the cobalt variant exhibits a lower growth rate and a higher condition factor in comparison with normal trout among other detrimental effects (Yamazaki 1974, Oguri 1976, Yada et al. 2002). Therefore, given the variation in the overall performance of blue-colored morphs of rainbow trout reported to date, specific assays are required to identify the possible existence of drawbacks, prior to considering a particular blue-colored strain for culture.

The Blue Back strain of rainbow trout is an interesting blue-colored phenotype for farming purposes in Chile, since its silvery blue color along with the scarce spottiness and less intense red lateral band are desirable traits for consumers in certain international markets (Colihueque et al. 2011). Therefore, this strain could be attractive for national producers to strengthen the trout farming industry, which has generated significant income for our country in recent years (Colihueque and Estay 2010). The application of a similar strategy has been cost effective in other commercial fishes, such as tilapia, in which the red strains have been cultured on an industrial scale given that it is more widely accepted than the wild-type strain (Hilsdorf et al. 2002). In rainbow trout some progress has been made with regard to this issue in recent years, particularly in Europe (Kause et al. 2003), where selected breeding based on scoring for various appearance traits has been implemented. The Blue Back strain has been characterised according to its skin color pattern at the juvenile and adult stages (Colihueque et al. 2011, Colihueque 2014). However, growth performance data for intensive culture has not been collected. Here, we examined several rainbow trout families reared under standard farming conditions for two years, in order to ascertain the effect of blue color intensity on growth performance. Furthermore, dark spots were counted and measured to provide insight into the effect of this phenotype on spottiness. This attribute is also important to be assessed given that this may influences consumer acceptance, since in some international market they prefer trout marketed whole with few dark spots on the skin (Colihueque et al. 2011). National producers may find this information useful for improving the external appearance of farmed rainbow trout.

In the current study we analysed the blue color intensity of three groups of the Blue Back strain of rainbow trout reared under standard farming conditions, to determine whether the variation of this color attribute could affect body weight and skin spottiness. To address this aim, first, we measured the blue color on the back by means of computer-based image analysis using spectral analysis software, based on images taken 282 days post fertilisation, since at this stage the full expression of this color is reached. Secondly, individuals were categorised into two contrasting classes within each group, namely, low and high blue color intensity, in order to determine, by comparison, the effect of the blue color variation on the body weight and skin spottiness of fishes. This design allowed us to determine the importance of the blue color variation on growth performance and external appearance, i.e., skin spottiness, both highly important traits of farmed rainbow trout.

2. Material and methods

2.1 Blue-colored groups of rainbow trout

Three independent groups of rainbow trout, PBBPR09 (n = 200), PBBIN07 (n = 138) and PBBOT07 (n = 243), were analysed. These groups belong to selection lines derived from a selection program for the Blue Back trait that was undertaken in the Huililco Ltd fish farm located in the Pucón district, La Araucanía Region, Chile, and were composed from 25 to 37 families. These families were composed of full-sib and half-sib progenies obtained by mating one male breeder with two or three female breeders, which were produced during winter 2007 (PBBIN07), autumn 2007 (PBBOT07) and spring 2009 (PBBPR09). Within each group, families had different ages due to spawning time differences among the female breeders used for crossbreeding at the time of the constitution of families. A random sample of fish per family was tagged individually when they had achieved a body weight of 5.0 g with passive integrated transponders (PIT) for their unambiguous identification to be subsequently cultured as a pool at a temperature of 9–11°C, with a constant water flow at 7 L/s and a culture density of 1–20 kg/m^3 in a 30 m^3 raceway pond with a green background. During rearing, fish were fed with extruded food free of pigment (BIOMAR, Golden Trucha) and composed of 47% protein, 20% lipid, 10% ash, and 1% crude fiber at 1.5% of live body weight per day. Pellet size was increased, according to the manufacturer's recommendations as the fish grew. At the end of the rearing periods, individually PIT-tagged fish from each family of each group were measured for body weight and total length to the nearest 0.1 g and 0.1 cm, respectively. For PBBOT07 group both biometric parameters were registered at the beginning of rearing period to calculate specific growth performance parameters. Fish were anesthetised with 100 mg/L of benzocaine solution before handling.

2.2 Image analysis

Rainbow trout individuals were photographed at the end of the rearing period to record their skin coloration and spottiness. At this stage (age-1+ or age-2+ depending group) full expression of the blue color and spottiness of the skin for this species is reached (Colihueque 2014), and therefore, registering the skin pigmentation parameters at this time was considered appropriate. Moreover, blue color intensity of the skin at adult stages in rainbow trout must reflect its expression at early development stages, due to this character has a full and constant expression at least after 282 days post fertilisation (Colihueque 2014). Thus, the analysis of possible effects of this condition on different growth performance parameters would be appropriate at this development stage. The image analysis involved measurements of skin blue color intensity on the back (BCI), the number of dark spots and dark spot size, as described by (Colihueque et al. 2011). Briefly, this method involved photographing of the entire trout from the dorsal view (from back to lateral line) using a 7.0 Mpx digital camera and lighting system with color temperature adjustment. Then, the photographs were stored at 24-bit RGB color depth on a memory card and transferred to a computer for color analysis. The BCI was defined as the percentage of skin area covered by blue/sky-blue hues [(number of matching pixels of blue/sky-blue hues/total number of pixels) x 100] registered within the 3 x 3 cm area of the back skin located in front of the dorsal fin. This parameter was calculated using the color map function and color threshold tool of the SIGMASCAN PRO 5.0 software. Dark spots were measured in grayscale with IMAGE J 1.38 software using the particle analysis tool and expressed as the mean number of dark spots (NDS) per area (number of spots/cm^2) and the mean dark spot size (DSS) in the area analysed (cm^2). To better analyze the effects of the BCI on the growth performance and spottiness, each group were categorised in contrasting groups according to the inter-individual variation of the BCI on the back. Thus, two categories were defined: 1) fishes with BCI < 50% (Low-BCI) and 2) individuals with BCI ≥ 50% (High-BCI) (Fig. 1). Further, to analyze the effect in more contrasting categories, fishes of each group were also sorted into 1st (Low-BCI) and 4th (High-BCI) quartiles.

2.3 Growth performance evaluation

The growth performances of fish was evaluated by calculating the following parameters: 1) Specific growth rate (SGR, %/day) = $[(\ln BW_f - \ln BW_i)/D] \times 100$; 2) Condition factor (CF) at the end of the testing periods, CF = $(BW_f/TL^3) \times 100$; 3) Daily growth (DG, g) = $(BW_f - BW_i)/D$; 4) Relative growth (RG, %) = $[(BW_f - BW_i)/ BW_i] \times 100$; 5) Coefficient of Variation (CV, %) at initial and final body weight = (SD /BW) x 100; 6) Absolute body weight gain (AWG, g) = $(BW_f - BW_i)$; and 7) Relative body weight gain (RWG, %) = $(BW_f - BW_i) \times 100/ BW_i$; where, BW_f = final body weight (g), BW_i = initial body weight (g), D = number of days in the rearing (rearing time), and TL = final total length (cm). Calculations were done based on measurements of individually PIT-tagged fish from each family.

2.4 Statistical analysis

Pearson product-moment linear correlations were done to assess the relationship between the BCI and the growth performance and spottiness parameters in each group. To test the significant differences among means of the BCI categories (Low-BCI vs. High-BCI), the Student's t-test for independent samples were used. Prior to statistical analysis the variables were analysed for conformance to assumptions regarding normal distribution and homogeneity of variance using the Kolmogorov–Smirnov and Levene´s tests, respectively. All statistical analyses were performed using the STATISTICA 5.1 program.

3. Results

Summary statistics of body weight and skin color parameters of blue-colored groups of rainbow trout are presented in Table 1. The association analysis (Table 2) revealed that BCI was significantly and positively related to body weight in all groups (PBBOT07: $r = 0.1966$, $P = 0.0021$; PBBPR09: $r = 0.1545$, $P = 0.0289$; PBB07IN: $r = 0.1809$, $P = 0.0337$). Thus, more heavy blue colored individuals had higher body weight (Fig. 2). Since the age of individuals within groups was not significantly correlated to body weight (PBBOT07: $r = 0.1070$, $P = 0.0962$; PBBPR09: $r = 0.1275$, $P = 0.0721$; PBB07IN: $r = -0.0362$, $P = 0.6731$), age effects may not have inflated the relationship between the body weight and BCI. The correlation analysis of spottiness parameters indicated that there was a significant negative association between NDS and BCI, but in one group (PBBOT07: r = -0.23869, $P = 0.0002$; PBB07IN: r = 0.0435, $P = 0.6124$), whereas no significant association was observed among DSS with BCI in any group (PBBOT07: r = -0.0626, $P = 0.3312$; PBB07IN: r = 0.0833, $P = 0.3311$).

The analyses of categorised data by BCI in PBBOT07 and PBB07IN groups are presented in Table 3. In both group, we noted a higher body weight in individuals of BCI ≥ 50% than in BCI < 50%, although this difference was significant only in group PBBOT07 ($P = 0.0459$) (Fig. 3a). In the quartile analysis, a more clear trend was observed, since body weight was significantly higher in the 4th than in the 1st quartile of BCI in both groups (PBBOT07, $P = 0.0459$; PBB07IN, $P = 0.0482$) (Fig. 3b). In the spottiness analysis, NDS was significantly lower in High-BCI (BCI ≥ 50% and 4th quartile,) than in Low-BCI (BCI < 50% and 1st quartile) categories, although this trend was significant for PBBOT07 group only ($P = 0.0014$ and $P = 0.0004$, respectively). DSS parameter did not present significant differences among BCI categories in any group.

The analysis of growth performance (Table 4) revealed a slightly higher body weight gain in individuals of High-BCI (BCI ≥ 50%) compared to those of Low-BCI (BCI < 50%). Thus, body weight gain was +11.76 g (8.73%) in PBBOT09, +20.95 g (4.77%) in PBBIN07, and +21.05 g (5.66%) in PBBOT07. However, these differences were only statistically significant in the PBBOT07 group ($P = 0.0459$). Further analysis of growth rates in PBBOT07 group, where a more complete analysis of performance parameters were carried out at 910 days (Table 5), revealed similar means for TL, SGR and CF among both categories of BCI, but with a statistically significant difference of means for DG ($P = 0.0446$), with the High-BCI (BCI ≥ 50%) category presenting highest DG value than those of Low-BCI (BCI < 50%) category (0.54 ± 0.10 and 0.51 ± 0.09, respectively). Thus, the better performing of individuals of the High-BCI category may be attributable mainly upon a slightly increasing (+5.88%) of mean daily growth.

4. Discussion

Several studies reveals that color morphs of rainbow trout may be associated with negative pleiotropic effects on growth performance parameters, represented by slower growth than wild-types, such as have been reported for albino (Dobosz et al. 2000, Okumus et al. 2001, Ercen et al. 2007), palomino (Dobosz et al. 2000) and golden (Clark 1970) phenotypes, including some blue-colored variants (Yamazaki 1974, Blanc et al. 2006, Oguri 1976). In our case, given that we found a significantly positively relationship between blue color intensity of the skin and the body weight in all blue-colored groups analysed, it seem that this condition, contrary to what one might suppose, produces a positive effect on the growth performance in rainbow trout. This trend was established by comparing Low-BCI and High-BCI categories of blue color intensity. Indeed, High-BCI fish (BCI $\geq$ 50%) outperformed the Low-BCI fish (BCI < 50%) for body weight in most groups, reaching a relative body weight gain above 4.77%. These results add new evidence on positive effect of the blue-colored phenotype on body weight in rainbow trout, an effect that was first reported by Kincaid (1975) in a blue-colored variant of this species from the United States, called iridescent metallic blue, a phenomenon that was revealed by a higher growth rate of the blue-colored morph (18–27%) than normal colored trout.

From a commercial point of view, the positive correlation between the blue color intensity of the skin and the growth rate in blue-colored individuals suggest that this phenotype is highly suitable for farming, since the appealing external coloration of this phenotype, combined with its higher growth rate, should increase the profitability of culture. Currently, this situation is not minor, given that rainbow trout farming in Chile, depend on the continuous improvement of a set of productive parameters to improve its profitability, where the external appearance constitutes one more of those characters (Kause et al. 2003, Colihueque 2010).

However, more investigation is required to discard the effect of factors that may have affected our results. To address this problem, new experimental design should be implemented, for example, to discard the social effects that may have contributed to the differences in growth between High-BCI fish and Low-BCI fish. In salmonid species it has been demonstrated that this effect may yield to an unavoidable social hierarchy when specimens of the same species are reared in small units, as occurred in intensive culture and when continuous access to feed is not possible, leading to competition for food (Okumus et al. 2001, Jobling 1985, Holm et al. 1990). Under such conditions, dominant fish are more active and consume most of the food that may be the case of High-BCI fish, while subordinate fish have a more restricted feed consumption. Thus, separation as opposed to communal rearing implemented in this study, should be analysed in future assays to rule out this effect. In addition, the possible occurrence of physiological and morphological color change should also be considered due as these phenomena usually influence variation in skin coloration of fishes in response to environmental factors (Sugimoto 2002), an effect that leads to phenotypic plasticity. For example, in brown trout populations the intensity of fish coloration, at least for melanin-based coloration, may vary significantly to match their background color (Westley 2013). In other species, variation in skin coloration has also been reported in response to density (Metusalach et al. 1997) and light (Donnelly and Whoriskey 1991) variations. Although the existence of these types of effects has not yet been proven in blue-colored phenotypes of rainbow trout, this cannot be totally discarded as available data of other blue-colored fishes indicate that the blue color expression of the skin may be subjected to physiological color change (Goda and Fujii 1998, Amiri and Shaheen 2012). Stress (Green et al. 1991, Iger et al. 1995) and management environment (Fleming et al. 1994) may influence this type of color change, a factors that is difficult to control in fish farming.

Another interesting finding from this study, though for at least one group, is the negative relationship found between the blue color intensity and dark spot number in the blue-colored phenotype. This result agrees with previous study carried out at the juvenile stage in a similar blue-colored phenotype of rainbow trout, in which a slight negative correlation between the two parameters was also observed (Colihueque 2014). Therefore, these data suggest an association between reduced skin spottiness of the blue-colored rainbow trout as a function of the blue color intensity. This result would be consistent with what has been observed in other blue-colored phenotypes rainbow trout, such as the cobalt variant, which shows a depigmented skin attributable to a reduction in dermal melanophores (Oguri 1983).

The covariation between the blue coloration and other phenotypic traits, such as the growth performance and the spottiness parameters, suggests a pleiotropic effect for the genes underlying this phenotype. To date, remains unknown which gene(s) control this trait in rainbow trout. Although it was not addressed in this study, candidate genes may include those that control pteridin and melanin synthesis based on biochemical analysis of cobalt-variants specimens of this species, which show significantly reduced levels of these pigments as compared to typical trout (Yamaguchi and Miki 1981). In addition, given that the blue coloration in fish is generally considered a structural color rather than a true color, emerging through the Tyndall effect, genes controlling the structural changes to skin layers may also be involved. In fact, evidence obtained in other fish indicates that some genes related with the structure of the skin appear to participate in the origin of blue coloration. For example, in cichlids the blue color appears to arise from a differential expression of genes related to collagen synthesis (Gunter et al. 2011). Further studies, ideally using a combination of molecular genetics and experiments focusing on specific gene interactions with phenotypic traits, would be helpful in analysing the maintenance or disruption of the correlation between the blue-colored phenotypes and growth performance or spottiness. This information will better elucidate the variations in this interaction. For example, some blue-colored phenotypes of common carp (Wlodek 1968, Wohlfarth and Moav 1970) produce opposing effects on growth performance, which suggests different underlying genetic bases for this trait. To date the analysis of the genetic basis of blue-colored phenotypes of rainbow trout, such as the new blue variant (Blanc et al. 2006) and the Blue Back strain (Diaz et al. 2011), indicates a different underlying genetic for these phenotypes.

5. Conclusion

This study indicates that the intensity of the blue color on the back skin of a blue-colored phenotype of rainbow trout, is positively correlated with growth performance. In addition, there would be an inverse relationship between blue skin color intensity and number of dark spot number; that is, fish with greater blue color intensity tend to have skin that is less spotty. These findings may be useful to strengthen trout farming in Chile, given that these characteristics are highly desirable to increase the profitability of this culture.

6. Acknowledgements

This study was partially financed by the Innova-Corporación de Fomento de la Producción de Chile (CORFO) Project 05CT6PPT–08.

7. References

Amiri MH, Shaheen HM. Chromatophores and color revelation in the blue variant of the Siamese fighting fish (*Betta splendens*). Micron 2012;43:159–69. [http://dx.doi.org/10.1016/j.micron.2011.07.002]

Blanc JM, Poisson H, Quillet E. A blue variant in the rainbow trout, *Oncorhynchus mykiss* Walbaum. J Hered 2006;97:89–93. [http://dx.doi.org/10.1093/jhered/esj010]

Clark FH. Pleiotropic effects of the gene for golden color in rainbow trout. J Hered 1970;61:8–10.

Colihueque N, Estay FJ. Trout farming in Chile. Country leads global production, further growth possible. Global Aquaculture Advocate Magazine 2010, January/February:30–1. Available from: https://www.researchgate.net/publication/307578021

Colihueque N, Parraguez M, Estay FJ, Diaz NF. Skin color characterization in rainbow trout by use of computer-based image analysis. N Am J Aquacult 2011;73:249–58. [http://dx.doi.org/10.1080/15222055.2011.581578]

Colihueque N. Analysis of the coloration and spottiness of Blue Back rainbow trout at a juvenile stage. J Appl Anim Res 2014;42:474–80. [http://dx.doi.org/10.1080/09712119.2014.883317]

Colihueque N. Genetics of salmonid skin pigmentation: clues and prospects for improving the external appearance of farmed salmonids. Rev Fish Biol Fisher 2010;20:71–86. [http://dx.doi.org/10.1007/s11160-009-9121-6]

Diaz NF, Estay F, Araneda C, Colihueque N. Heritability of Blue Back skin colour and correlations with growth in rainbow trout, *Oncorhynchus mykiss*, 2011. Proceedings of the Australian Society for Fish Biology Conference: Townsville, Australia 2011; 72.

Dobosz S, Kohlmann K, Goryczko K, Kuzminski H. Growth and vitality in yellow forms of rainbow trout. J Appl Ichthyol 2000;16:117–20. [http://dx.doi.org/10.1046/j.1439-0426.2000.00147.x]

Donnelly WA, Whoriskey FG. Background-Color Acclimation of Brook Trout for Crypsis Reduces Risk of Predation by Hooded Mergansers *Lophodytes cucullatus*. N Am J Fish Manage 1991;11:206–11. [http://dx.doi.org/10.1577/1548-8675(1991)011<0206:BCAOBT>2.3.CO;2]

Ercen Z, Tekelioglu N, Ozluer A, Unaldi E, Ozkan F. Comparasions of growth performance of albino and normal pigmented rainbow trout (*Oncorhynchus mykiss* Walbaum, 1792). Turkish Journal of Aquatic Life 2007;3–5:738–46.

Fleming IA, Jonsson B, Gross MR. Phenotypic divergence of sea-ranched, farmed, and wild salmon. Can J Fish Aquat Sci 1994;51:2808–24. [http://dx.doi.org/10.1139/f94-280]

Goda M, Fujii R. The blue coloration of the common surgeonfish, *Paracanthurus hepatus*—II. Color revelation and color changes. Zool Sci 1998;15:323–33. [http://dx.doi.org/10.2108/zsj.15.323]

Green JA, Baker BI, Kawauchi H. The effect of rearing rainbow trout on black or white backgrounds on their secretion of melanin-concentrating hormone and their sensitivity to stress. J Endocrinol 1991;128 :267–74. [http://dx.doi.org/10.1677/joe.0.1280267]

Gunter HM, Clabaut C, Salzburger W, Meyer A. Identification and characterization of gene expression involved in the coloration of cichlid fish using microarray and qRT-PCR approaches. J Mol Evol 2011;72:127–37. [http://dx.doi.org/10.1007/s00239-011-9431-x]

Hilsdorf AWS, Penman DJ, Farias EC, Macandrew B. Melanophore appearance in wild and Red Tilapia embryos. Pigm Cell Res 2002;15:57–61. [http://dx.doi.org/10.1034/j.1600-0749.2002.00058.x]

Holm JC, Refstie T, Bø S. The effect of fish density and feeding regimes on individual growth rate and mortality in rainbow trout (*Oncorhynchus mykiss*). Aquaculture 1990;89:225–32. [http://dx.doi.org/10.1016/0044-8486(90)90128-A]

Iger Y, Balm PHM, Jenner HA, Bonga SEW. Cortisol induces stress-related changes in the skin of rainbow trout (*Oncorhynchus mykiss*). Gen Comp Endocr 1995;97:188–98. [http://dx.doi.org/10.1006/gcen.1995.1018]

Jobling M. Physiological and social constraints on growth of fish with special reference to Arctic charr, *Salvelinus alpinus* L. Aquaculture 1985;44:83–90. [http://dx.doi.org/10.1016/0044-8486(85)90011-0]

Kause A, Ritola O, Paannanen T, Eskelinen U, Mantysaari E. Big and beautiful? Quantitative genetic parameters for appearance of large rainbow trout. J Fish Biol 2003;62:610–22. [http://dx.doi.org/10.1046/j.0022-1112.2003.00051.x]

Kincaid HL. Iridiscent metallic blue color variant in rainbow trout. J Hered 1975;66:100–2.

Metusalach, Brown JA, Shahidi F. Effects of stocking density on colour characteristics and deposition of carotenoids in cultured Arctic charr (*Salvelinus alpinus*). Food Chem 1997;59:107–14. [http://dx.doi.org/10.1016/S0308-8146(96)00205-1]

Oguri M. Dermal melanophores and renal melanin-containing cells in the "cobalt" variant of rainbow trout. Bull Jpn Soc Sci Fish 1983;49:1679–81.

Oguri M. On the enlarged liver in "cobalt" variant of rainbow trout. Bull Jpn Soc Sci Fish 1976;42:823–30.

Okumus I, Degirmenci A, Bascınar N, Celikkale MS. Comparative performance, approximate biochemical composition and consumer preference of albino and normally pigmented varieties of rainbow trout (*Oncorhynchus mykiss*). Turk J Fish Aquat Sc 2001;1:23–8.

Sugimoto M. Morphological color changes in fish: Regulation of pigment cell density and morphology. Microsc Res Techniq 2002;58:496–503. [http://dx.doi.org/10.1002/jemt.10168]

Westley PAH, Stanley R, Fleming IA. Experimental tests for heritable morphological color plasticity in non-native brown trout (*Salmo trutta*) populations. PloS One 2013;8:e8040. [http://dx.doi.org/10.1371/journal.pone.0080401]

Wlodek JM. Studies on the breeding of carp (*Cyprinus carpio*) at the experimental pond farms of the Polish Academy of Sciences in southern Silesia, Poland. World Symposium on Warm-water Pond Fish Culture, Rome: FAO Fish Report No.44.; 1968, p. 93–116.

Wohlfarth G, Moav R. The effects of variation in spawning time on subsequent relative growth rate and viability in carp. Bamidgeh 1970;22:42–7.

Yada T, Moriyama S, Suzuki Y, Azuma T. Relationships between obesity and metabolic hormones in the "cobalt" variant of rainbow trout. Gen Comp Endocr 2002;128:36–43. [http://dx.doi.org/10.1016/S0016-6480(02)00047-3]

Yamaguchi K, Miki W. Comparison of pigments in the integument of cobalt, albino, and normal rainbow trout, *Salmo gairdnerii irideus*. Comp Biochem Phys B 1981;68B:517–20.

Yamazaki F. On the so-called "cobalt" variant of rainbow trout. Bull Jpn Soc Sci Fish 1974;40:17–25.

Table 1. Summary statistics of body weight and skin color parameters of blue-colored groups of rainbow trout.

Groups	No. of individuals	No. of families	Age range (days)	Body weight (g)	Blue color intensity on the back (% of area covered by color)	Number of dark spots (No. of spots/cm^2)	Dark spots size (cm^2)
PBBPR09	200	37	403-467	145.31 ± 33.53 (56-254) [0.23]	87.95 ± 22.74 (1.11-100) [0.26]	-	-
PBBIN07	138	34	940-998	447.60 ± 64.94 (330-631) [0.15]	46.74 ± 31.68 (0.07-100) [0.68]	4.09 ± 1.61 (0.00-8.00) [0.39]	2.91 ± 1.55 (0.00-8.55) [0.53]
PBBOT07	243	25	871-910	376.45 ± 67.51 (208-564) [0.18]	28.20 ± 28.34 (0.00-98.59) [1.00]	1.45 ± 0.80 (0.00-3.78) [0.55]	0.07 ± 0.03 (0.00-0.18) [0.46]

Values are mean ± standard deviation of mean. Range in round parenthesis. Coefficient of variation in squared parenthesis

Table 2. Simple linear correlations between the blue color intensity and the growth performance and spottiness parameters in blue-colored groups of rainbow trout.

Correlation	PBBOT07 (n = 243)	PBBPR09 (n = 200)	PBB07IN (n = 138)
BCI vs. BW	0.1966 (0.0021)**	0.1545 (0.0289)*	0.1809 (0.0337)*
BW vs. Age	0.1070 (0.0962)^{NS}	0.1275 (0.0721)^{NS}	-0.0362 (0.6731)^{NS}
BCI vs. NDS	-0.23869 (0.0002)***	-	0.0435 (0.6124)^{NS}
BCI vs. DSS	-0.0626 (0.3312)^{NS}	-	0.0833 (0.3311)^{NS}

*Number (n) of individuals from each studied groups. Bold font denotes significant correlations (*P < 0.05, **P < 0.01, ***P < 0.001) according to Student's t-test. The exact P-values are indicated in parenthesis. BCI = blue color intensity on the back, BW = body weight, NDS = number of dark spots, DSS = dark spots size*

Table 3. Comparison for spottiness parameters of blue-colored groups of rainbow trout categorised according to the blue back color intensity on the back.

Groups	Low-BCI (BCI < 50%)	High-BCI (BCI ≥ 50%)	Student`s t-test (Exact p-value)
PBBIN07			
n	80	58	
BCI value (%)	23.74 ± 16.11	78.46 ± 16.91	
BW (g) NS	438.79 ± 64.17	459.74 ± 64.57	0.0612
NDS (N° of dark spots/cm^2) NS	4.10 ± 1.46	4.07 ± 1.81	0.9182
DSS (cm^2) NS	2.90 ± 1.50	2.91 ± 1.64	0.9794
PBBOT07			
n	191	52	
BCI values (%)	15.45 ± 14.63	75.05 ± 12.91	
BW (g) *	371.95 ± 65.81	393.00 ± 71.65	0.0459
NDS (N° of dark spots/cm^2) **	1.54 ± 0.79	1.14 ± 0.76	0.0014
DSS (cm^2) NS	0.07 ± 0.03	0.06 ± 0.03	0.4517
	1st quartile	4th quartile	
PBBIN07			
n	38	35	
Quartile values (BCI)	9.08 ± 7.39	90.66 ± 7.95	
BW (g) *	426.57 ± 56.79	453.56 ± 57.87	0.0482
NDS (N° of dark spots/cm^2) NS	4.13 ± 1.55	4.27 ± 1.93	0.7333
DSS (cm^2) NS	2.88 ±1.49	3.15 ± 1.72	0.4703
PBBOT07			
n	62	61	
Quartile values (BCI)	1.41 ± 1.28	70.92 ± 15.56	
BW (g) **	359.55 ± 59.99	391.26 ± 70.21	0.0080
NDS (N° of dark spots/cm^2) ***	1.69 ± 0.79	1.18 ± 0.75	0.0004
DSS (cm^2) NS	0.07 ± 0.03	0.06 ± 0.03	0.2869

*Values are mean ± standard deviation of mean. n = number of individuals, BCI = Blue color intensity on the back, BW = body weight, NDS = number of dark spots, DSS = dark spots size. *P < 0.05; **P < 0.01, ***P < 0.001, NS = not significant.*

Table 4. Comparison for growth performance parameters of blue-colored groups of rainbow trout categorised according to the blue back color intensity on the back.

	PBBOT07			PBBOT09			PBBIN07		
Parameter	Low-BCI (BCI ≤ 50%)	High-BCI (BCI > 50%)	Student`s t-test (Exact p-value)	Low-BCI (BCI ≤ 50%)	High-BCI (BCI > 50%)	Student`s t-test (Exact p-value)	Low-BCI (BCI ≤ 50%)	High-BCI (BCI > 50%)	Student`s t-test (Exact p-value)
n	191	52		20	180		80	58	
BCI value (%)	15.45 ± 14.63	75.05 ± 12.91		27.83 ± 13.82	94.62 ± 10.29		23.74 ± 16.11	78.46 ± 16.91	
BW (g)	371.95 ± 65.81	393.00 ± 71.65	0.0459 *	134.73 ± 34.86	146.48 ± 33.27	0.1372 ^{NS}	438.79 ± 64.17	459.74 ± 64.57	0.0612 ^{NS}
AWG (g)	+21.05			+11.76			+20.95		
RWG (%)	+5.66			+8.73			+4.77		

*Values are mean ± standard deviation of mean. n = number of individuals, BCI = Blue color intensity on the back, BW = body weight, AWG = absolute body weight gain, RWG = relative body weight gain. *P < 0.05; NS = not significant.*

Table 5. Comparison for growth performance parameters of the blue-colored PBBOT07 group of rainbow trout categorised according to the blue back intensity of color on the back.

Parameter	Low-BCI (BCI < 50%)	High-BCI (BCI ≥ 50%)	Student's t-test (Exact p-value)
n	191	52	
BW_i (g) ***	5.62 ± 1.81	6.64 ± 1.81	0.0002
BW_f (g) *	371.95 ± 65.81	393.00 ± 71.65	0.0459
TL (cm) ^{NS}	31.54 ± 1.96	32.13 ± 1.80	0.0523
SGR (%/day) ^{NS}	5.90 ± 0.18	5.95 ± 0.19	0.0626
CF ^{NS}	1.18 ± 0.11	1.17 ± 0.10	0.7595
DG (g) *	0.51 ± 0.09	0.54 ± 0.10	0.0446
RG (%) *	7204.11 ± 2591.81	6258.52 ± 2108.74	0.0162

*Values are mean ± standard deviation of mean. n = number of individuals, BCI = Blue color intensity on the back, BW_i= initial body weight, BW_f = final body weight; TL = final total length, SGR = specific growth rate, CF = condition factor, DG = daily growth, RG = relative growth. *$P < 0.05$; ***$P < 0.001$; NS = not significant*

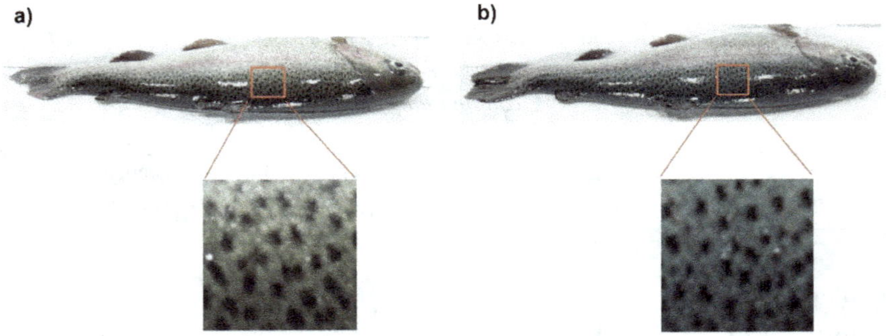

Fig. 1 Images show two specimens of the blue-colored phenotype of rainbow trout, sorted in contrasting color categories based on blue color intensity on the back (BCI). a) Specimen of Low-BCI category (BCI < 50%) and b) individual of High-BCI category (BCI ≥ 50%). The BCI was defined as the percentage of skin area covered by blue/sky-blue hues [(number of matching pixels of blue/sky-blue hues/total number of pixels) x 100] registered within the 3 x 3 cm area of the back skin located in front of the dorsal fin. The region of interest is represented by an expanded rectangle.

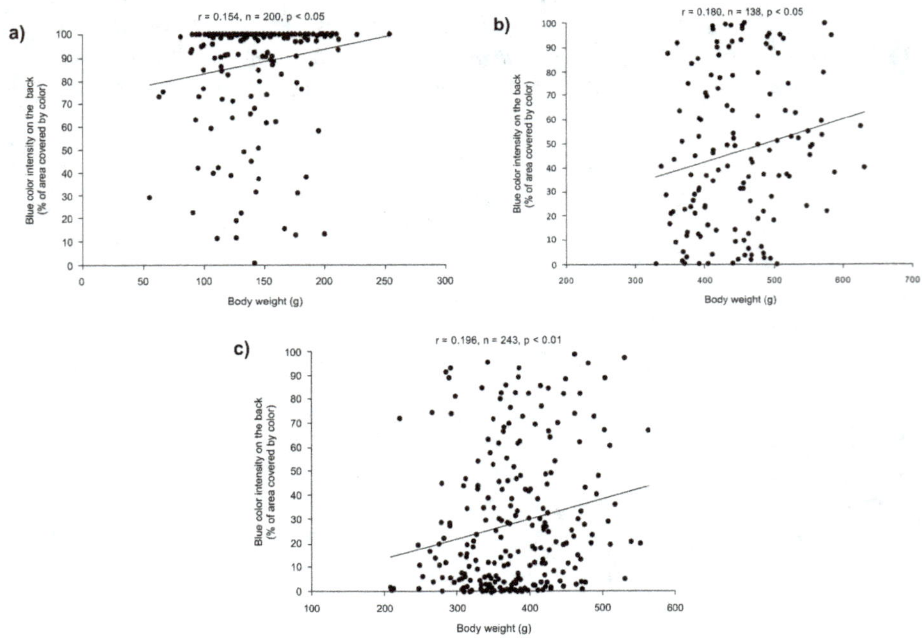

Fig. 2 Pearson product-moment linear correlations between the blue color intensity on the back and the body weight in three blue-colored groups of rainbow trout. a) PBBPR09 group (n = 200), b) PBBIN07 group (n = 138) and c) PBBOT07 group (n = 243). The regression line is indicated in a black line.

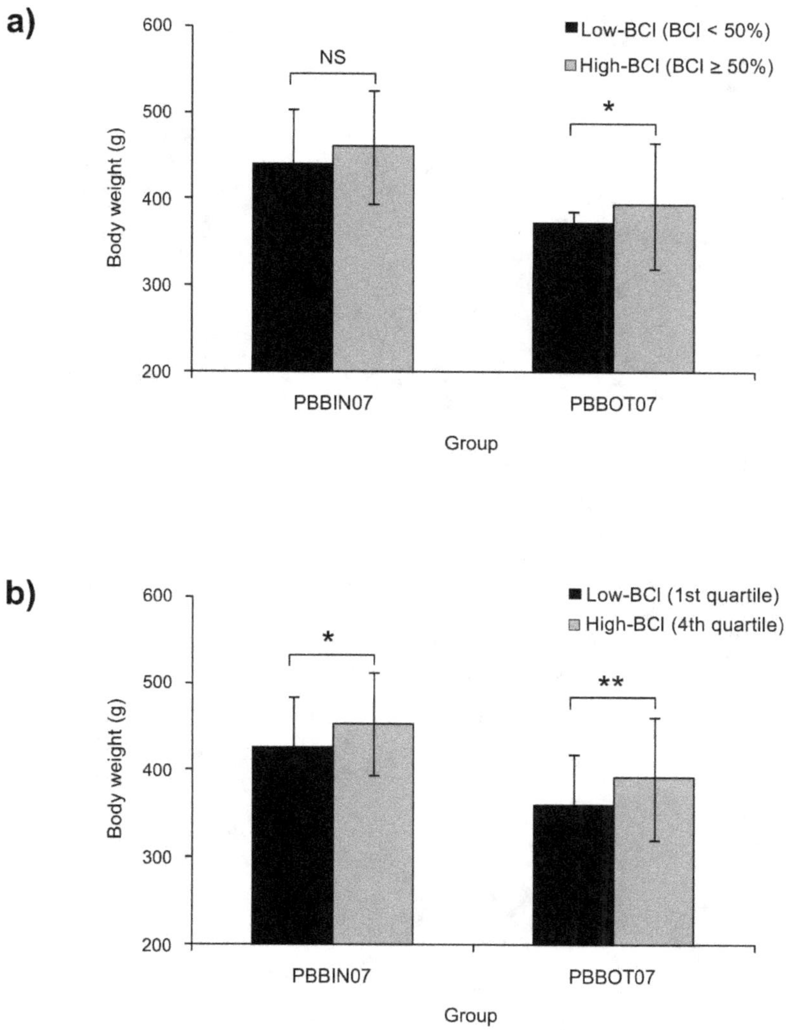

Fig. 3 Comparison of growth performance of blue-colored groups of rainbow trout. Individuals of each group were sorted on contrasting categories base on blue color intensity on the back (BCI). a) comparison of body weight among BCI < 50% and BCI ≥ 50% categories, and b) comparison of body weight between 1st and 4th quartiles. The BCI < 50% and 1st quartile were considered a Low-BCI category, whereas BCI ≥ 50% and 4th quartile corresponded to the High-BCI category. Columns and bar errors represent means and standard deviation of means, respectively. Significant differences between means were calculated according to Student's t-test. *P < 0.05, **P < 0.01, NS = not significant.

Chapter 5.

Photo-identification of rainbow trout *Oncorhynchus mykiss* (Walbaum, 1792) individuals using the flank spot pattern

Abstract

The flank skin spot patterns for identification of rainbow trout individuals may be useful as an alternative non-invasive tagging method for addressing biological questions related to this species, including issues relevant to fish farming. This approach has potential because the spot arrangement is usually specific to each individual. In this study, we analysed the flank spot patterns of individuals from two rainbow trout *Oncorhynchus mykiss* (Walbaum, 1792) populations—one naturalised population (n= 56, Calafquén Lake, La Araucanía Region) and one farmed population (n= 45, Piscícola Huililco Ltd., La Araucanía Region)—to evaluate whether these spot patterns are useful for the photo-identification of fish. The flanks of the individuals were photographed with a digital camera. Then, the spot pattern of each image was annotated with freely available software (Interactive Individual Identification System, I3S) to evaluate the performance of these patterns in identifying individuals based on a ranking image score where each image was compared with all annotated images stored in an identification database. Spot patterns for long-term identification were also evaluated using a subset of images from the farmed population. The use of spot patterns showed a high performance for the identification of individuals, because in both populations, all annotated images ranked the first ranking score, indicating perfect recognition. The long-term analysis revealed a high performance in recognition of individuals because 75% of the images of fish at 2-year-old were successfully matched with images of the same fish at 3-year-old.

Key words

Individual recognition, spot pattern, non-invasive tagging, matching algorithms, fish farming

Note: Reproduced from Colihueque and Estay (2025). Acta Ichthyologica et Piscatoria 55, 181–190, DOI: 10.3897/aiep.55.151044 (CC BY 4.0)

1. Introduction

The use of skin pigmentation as an identification tool has been explored in several fish, given that body pigmentation can vary across individuals, sexes, species, geographic areas, or origins (e.g., native versus hatchery) (Katoh et al. 2001; Aparicio et al. 2005; Pavlidis et al. 2006; Luncentini et al. 2011; Martin-Smith 2011; Joo et al. 2013; Hustler 2014). In salmonids, previous investigations indicate the favourable use of skin pigmentation for identification of individuals (García de Leaniz et al. 1994; Merz et al. 2012; Stien et al. 2017; Cisar et al. 2021). For example, the size and position of the opercular spots in post-smolt Atlantic salmon (Stien et al. 2017), and the spot patterns on the head region of Chinook salmon (Merz et al. 2012), represent characteristics that can be used to identify individuals with high accuracy.

Although rainbow trout *Oncorhynchus mykiss* (Walbaum, 1792) present characteristic skin pigmentation that is unique within salmonids (Ade 1989), this pigmentation can display wide intraspecific variation in terms of colour, spottiness, and darkness (Islam et al. 1973; Tack 1973; Kause et al. 2003; Colihueque 2010; Colihueque et al. 2011, 2014, 2018). Thus, skin spottiness, represented by dark spots, may vary in number both on the flank (Islam et al. 1973; Tack 1973; Colihueque et al. 2018) and on the back (Colihueque et al. 2011, 2014) within and between populations. In some cultured stocks of this species, the spottiness variation can range from individuals with little or no spotting (< 0.5 spots/cm^2) to others with highly spotted skin (> 5.5 spots/cm^2) (Colihueque et al. 2011). To date, the origin of this variation in rainbow trout is unclear; however, some evidence indicates that the number of dark spots below the lateral line varies as a function of fish size and spawning time (Islam et al. 1973) and that variation in these spots on the back is associated with the age of juvenile fish (Colihueque et al. 2011). Other evidence indicates that the variation in the number of dark spots in this species, either above or below the lateral line, is controlled by genetic factors because the estimation of quantitative genetic parameters reveals a medium-low heritability value for this characteristic ($h^2 = 0.16$-0.61) (Kause et al. 2003; Díaz et al. 2011).

In rainbow trout, skin spottiness has been analysed from different points of view. Studies that have analysed the variation in the number of dark spots relative to different back colour group (Colihueque et al. 2011), their association with biometric and reproductive parameters (Colihueque et al. 2018), and their association with the blue colour intensity on the back skin (Colihueque and Estay 2018); additionally, other studies have assessed the genetic correlation of skin spottiness with body weight at harvest in cultured stocks (Rodriguez et al. 2019), and estimated the heritability of skin spottiness for the genetic improvement of the external appearance of fish (Kause et al.

2003). Little attention, however, has been given to the potential use of the skin spottiness in this species as a characteristic for the identification of individuals.

The photo-identification method involves the use of natural markings of any type that are present in individuals (e.g., fin shape and pigmentation patterning); these markings are recorded by photography and stored in a database for subsequent cross-matching analysis (Speed et al. 2007). This database can be examined manually to develop a set of individual matches (Meekan et al. 2006). However, when the number of photographs in a database increases, the development of faster, automated techniques is needed to compare new photographs to those previously stored. To address this problem, several automated matching algorithms have been developed for different animals with variable success rates (Hillman et al. 2003; Arzoumanian et al. 2005). In salmonids, the identification of individual fish has been addressed by different photo-identification methods. These include the visual examination of photographs to identify different constellations of melanophores by independent observers (García de Leaniz et al. 1994), the visual recognition of spot patterns combined with further point-pattern-matching algorithms based on star patterns (Stien et al. 2017), and the design of fully automated matching algorithms to score images of individuals on the basis of spot pattern recognition (Merz et al. 2012; Cisar et al. 2021). Thus, most photo-identification methods trialled in salmonids have examined the spot pattern, and specific matching algorithms have been used to improve image processing. There are no reports on the use of simple, freely available software that recognises spot patterns, such as Interactive Individual Identification System (I3S) (den Hartog and Reijns 2014), for use in photo-identification studies in salmonids.

In this study, we analysed the flank spot patterns of individuals from two rainbow trout populations—one naturalised population and one farmed population—to evaluate whether the spot pattern of each individual can be used for the photo-identification of fish. Using particular statistics, we first measured the regularity of the arrangement of skin spots. We next evaluated how well the individuals could be unambiguously identified using freely available software for individual fish recognition. By using this software, we were able to optimise this procedure by storing many annotated images in a database to facilitate the identification of individuals.

2. Methods

2.1 Fish used for the analysis

In 2007, adult naturalised rainbow trout individuals (n = 56; 39 females and 17 males) were collected from the inlet streams of Calafquén Lake (Villarrica district, La

Araucanía Region, Chile) during September-November, a period that cover the full spring run of this population. The collected fish were maintained in breeding ponds in Chiguaico Hatchery, which is located near Calafquén Lake, until the spawning phase, which was the moment when they were photographed. The cultured individuals (n = 45, 40 females and 5 males, 2 years old) were obtained from Piscícola Huililco Ltd. fish farm (Pucón district, La Araucanía Region, Chile), and the fish belong to the Blue Back strain reared in this centre (Colihueque and Estay 2018). All the individuals were tagged with electronic passive integrated transponder (PIT) tags to register their identification unequivocally during the photographing and the recognition processes. The transitory capture of the wild rainbow trout individuals from Calafquén Lake, were authorised by the Chilean Undersecretary of Fisheries and Aquaculture (permit #2750/19-10-2007). Fish management was in compliant with guidelines of the Chilean Undersecretary of Fisheries and Aquaculture related to the application of a protocol of sanitary management and animal care that must be applied in fish farms that operate under its jurisdiction (Decree DS319-2001).

2.2 Photography and spottiness regularity determination

The flanks of the individuals (left flank) were photographed to record the spot pattern of this area, represented by dark spots, using a 7.0 Mpx digital camera and natural lighting. The photographs were stored at 24-bit RGB colour depth on a memory card and transferred to a computer for spottiness analysis. The specimens were anaesthetised with benzocaine to immobilise them before the photographic sessions and were then immediately returned to the raceways for recovery. The rainbow trout were manually held to obtain a lateral view photograph of the entire live fish. The camera was held perpendicular to the plane of the rainbow trout's skin surface and positioned 80 cm away from the fish. No polarising filters were used. Photographs were taken on sunny days, between 11:00 and 16:00 hours, to avoid major changes in lighting conditions and to enhance the visualisation of the spot pattern in the area of interest. Under this photographing condition, however, some degree of specular reflections may have occurred because fish skin is a reflective surface that can obstruct features in the area of interest. Calafquén Lake individuals were photographed between October and November 2007 (spring season), while Blue Back strain individuals were recorded in February 2007 (summer season) and again in February 2008 for the long-term identification analysis. To analyse the spottiness regularity, the original colour image was transformed into an 8-bit greyscale image. The spottiness regularity for each image was analysed by selecting a 2 × 2 cm area from the photograph, particularly below the dorsal fin. The nearest-neighbour distances between spots were calculated by using the freely available ImageJ plugin FLEYE software v.10.4, which is a tool designed to measure regularity in biological patterns (Diez-Hermano et al. 2015). The spottiness regularity was determined by calculating the coefficients of variation for nearest-neighbour distances between spots (CV-NN), where a higher CV-NNs indicated low regularity of spot arrangement (i.e., the spot positions show high irregularity) (Parichy and Turner 2003). Before measuring of the spottiness regularity, a set of training data conforming to images with regular spot arrangements (n=29) was used to validate this process.

2.3 Individual recognition analysis

The fish images were classified into three categories (Table 1): 1) no dark spots (no visible or recognisable spots detected either with naked eye or with a computer algorithm); 2) well-formed dark spots (spots easily identifiable with naked eye or with a computer algorithm); and 3) diffuse dark spots or ambiguous spot patterns. For category 2 images, the identification of individual fish was assessed by using the I3S CLASSIC v.4.0.2 software (den Hartog and Reijns 2014), an algorithm that compares the spottiness of an image with other images stored in a database to search for similarities in spot pattern. This tool searches for similar spot positions among the images while considering the distances between spots by comparing two point clouds, resulting in an image score ranking. In this analysis, a score equal to zero (first ranking) indicates a complete match of the spot pattern between two images, indicating perfect recognition of the same individual. For this purpose, the software requires the selection of a triangle in the region of interest consisting of three fixed anatomical landmarks (i.e., reference points). In our case, these comprised the initial dorsal fin (first dorsal landmark), the end dorsal fin (second dorsal landmark), and the pelvic fin (third landmark) (Fig. 1). In addition, the location of the spots in the region of interest must be manually indicated to generate a fingerprint file that is stored in the subdirectory used in identification. The annotated images of an unknown animal are automatically matched with all annotated images in the identification database to yield a ranked list of images.

2.4 Evaluation of the spottiness pattern for long-term identification

A subset of images from the Blue Back strain (n = 8) was selected to determine if the flank skin spots present at 2-years-old were still present at 3-years-old. This task was conducted to evaluate whether the spot patterns could be used for long-term identification of rainbow trout. The I3S CLASSIC v.4.02 software was used to compare one individual's spot patterns at 2-years-old with spot patterns from multiple 3-years-old individuals. In this analysis, the first image ranking indicates a complete match of the spot pattern between two images, indicating perfect recognition of the individual. The complete matches of perfect recognition were subsequently verified using PIT tag information. The fish length and the number of spots present in the region of interest at the two developmental stages were also investigated.

2.5 Statistical analysis

Pearson correlation analyses were carried out to determine the relationship between the number of dark spots on the flank and the score rankings of the images. Student's t-test for independent samples was used to test the significance of differences among means of fish length and number of spots between the two developmental stages in the long-term identification analysis. Before statistical analysis, the variables were analysed to determine the conformance to assumptions regarding normal distribution and homogeneity of variance via the Kolmogorov-Smirnov test and Levene's test, respectively (Heumann et al. 2016). The statistical analyses were performed using the Statistica software version 5.1. Confidence intervals (CIs) for match probability of the recognition rate of the photo-identification method were calculated with a 95% CI for each population according to the procedure of Clopper and Pearson (1934) using R-Studio version 4.3.1 (R Studio Team 2017).

3. Results

The spottiness regularity did not differ significantly ($P > 0.05$) between the Calafquén Lake population (mean CV-NN = 0.58 ± 0.92) and the Blue Back strain (mean CV-NN = 0.73 ± 1.38) (Table 2). However, this parameter in these populations was significantly greater than the spottiness regularity of the training dataset (mean CV-NN = 0.12 ± 0.16) ($P < 0.05$). In addition, a large portion of the analysed individuals of both populations presented low regularity because at least 50% of them presented a value of CV-NN greater than or equal to 0.33 according to the Q2 quartile.

The individual recognition analysis indicated high performance for the identification of individuals in both populations because all annotated images ranked the first ranking score (mean ranking score = 0 ± 0) when they were compared with the annotated images stored in the identification database (Table 3 and Fig. 2). This result indicates that each analysed individual exhibited a distinctive spot pattern on the flank skin, allowing the perfect recognition of these individuals, i.e., with a 100% recognition rate in the Calafquén Lake population (95% CI = 93.1-100) and with a 100% recognition rate in the Blue Back strain (95% CI = 90.9-100) (Fig. 3). In contrast, large mean values were observed in the second and third ranking score images, from 7.02 to 7.76 in the Calafquén Lake sample, and from 8.28 to 9.16 in the Blue Back strain sample, supporting the absence of overlap with the first ranking score image. In addition, the second and third ranking score images showed a significant negative correlation ($P < 0.05$) with the number of spots in both populations (R = -0.75 and 0.78; R = 0.71 and 0.78, respectively) (Fig. 4), indicating a decrease in the ranking score as spot number on the flank skin increased.

The long-term identification analysis revealed high performance in the recognition of individuals because 6 out of the 8 images of 2-year-old fish were successfully matched with the images of the same fish at 3-year-old, as they ranked the first image ranking score, representing a 75% recognition rate (95% CI = 34.9-96.8) (Table 4). This result indicates that the spot pattern is highly consistent across different developmental stages (Fig. 5). The mean number of spots on the flank did not significantly differ between the two developmental stages ($P > 0.05$), revealing no trend for the mean number of spots to increase with fish growth ($P < 0.05$).

4. Discussion

The regularity analysis of this study indicates that rainbow trout individuals show a high level of irregularity in the spot pattern arrangement on the flank skin, indicating an uneven distribution of spot positions. Indeed, we found that at least 50% of the studied individuals had low regularity levels, with spot position values greater than or equal to 0.33 of the coefficient of variation. This result is consistent with previous investigations carried out in rainbow trout, which revealed strong irregularities in the spot positions on some body regions, particularly in the back skin (Colihueque 2014). Thus, this result provides evidence that the spot pattern in rainbow trout is a complex characteristic because it comprises variations in spot number (Islam et al. 1973; Tack 1973; Colihueque et al. 2014, 2018), spot size (Colihueque and Estay 2018), body surface distribution of the spot (Colihueque et al. 2011), and spot positions on the flank. Variation in the spot position appears not to be exclusive to rainbow trout because this type of variation has also been documented in other salmonids (Merz et al. 2012; Stien et al. 2017) and other fish (Persat 1982). For example, in Atlantic salmon opercular spots form a relatively specific pattern for each individual (Stien et al. 2017), and the spot pattern observed in the head region of Chinook salmon smolts presents similar characteristics (Merz et al. 2012). In contrast to these studies, however, we provided specific measures to quantify the degree of variation of this characteristic.

The results of this study indicate high performance in the recognition of individuals based on the spottiness of flank skin after a relatively large number of individuals were analysed using software-driven processing. The variability of the spot positions on flank skin was sufficient to correctly identify 100% of the analysed individuals of both populations. This result indicates that the spot pattern on flank skin is unique for each individual. The results of this study are similar to those reported in other salmonids that have used the spot pattern on the skin as a visual recognition tool. For example, in smolts of Chinook salmon, Merz et al. (2012) achieved 100% individual recognition (106 of 106 tested individuals) by using the spot pattern in the head region of fish with usable spot patterns. In this analysis, these authors reidentified fish from the previous period (approximately one month prior) through spot pattern recognition algorithms. In addition, Stien et al. (2017) used the size and position of opercular spots in post-smolt Atlantic salmon, and obtained an individual visual recognition success of 85% (17 of 20 tested individuals) by using a point-matching algorithm and comparing the images of 12-month-old fish images with those of the same fish at 22-month-old.

The photo-identification method tested in this study works well for fish with approximately ten to one hundred spots on the flank skin. It is unknown whether the photo-identification method maintains its recognition performance beyond this range. It is likely that a greater number of spots on the flank skin may reduce the performance of this method since a negative correlation between the second or third ranking score image and the number of spots on the flank skin was observed. Thus, images with a greater spot number could yield a ranking score image with values closer to zero (i.e., a barely perfect match), resulting in an incorrect match that may confuse the identification process. Moreover, the spots present on the skin of any analysed individual need to be well-formed for the generation of the fingerprint file to decrease the likelihood of annotating images with erroneous spot patterns. The present identification method is obviously not useful for the recognition of non-spotted fish. However, since these fish are observed at low frequencies in rainbow trout (Islam et al. 1973; Tack 1973; Colihueque et al. 2011, 2014, 2018; Colihueque and Estay 2018), as we confirmed in the present study (from 0 to 8.9%), such pigmentation conditions may not a major limitation in the use of this method for different rainbow trout populations.

The present photo-identification method has favourable prospects for the long-term identification of rainbow trout individuals because its high recognition rate (75%). However, since this result is based on a small number of individuals analysed, further examination with a larger sample size will be needed to better support this finding. In other salmonids, evaluation of the consistency of spottiness over time (10 months, 20 individuals) for the recognition of individuals show similar results (85%) when the arrangement of the opercular spots in post-smolt Atlantic salmon is considered (Stien et al. 2017). It is possible that growth-dependent changes in spot pattern on flanks, but not change in the spot number, could occur as a significant grow in length from 2-years-old to 3-years-old was observed during this period. This effect should be accounted for because changes in the spot position between developmental stages may decrease the performance of the present recognition method. Therefore, an algorithm that accounts for changes in the shape of the flank region during development is needed to improve the success of individual recognition at different developmental stages. Furthermore, sexual dimorphism bias in recognition success rate due to different spot pattern characteristics between the sexes is a factor that could have affected our results, given that an unequal proportion of males and females were analysed in the studied populations. However, this factor appears to be of less importance in this study because available data indicate that the spot pattern measured as the number of spots in the flank between males and females did not differ significantly, at least in the Calafquén Lake population (Colihueque et al. 2018). However, the strain skin colour effect between populations on the recognition rates cannot be discarded, because previous investigations indicate that rainbow trout individuals classified by their blue colour intensity (BCI) on the back (i.e., low-BCI and high-BCI) present significant differences in the number of dark spots in this body region (Colihueque and Estay 2018). Further support on spottiness differences between sexes in rainbow trout, but associated with the number of dark spots on the flank, can be observed when farmed individuals of this species are classified by the skin colour on their back in blue, yellow and green groups (Colihueque et al. 2011).

Other variables related to image collection, such as illumination and specular reflections, need to be considered to improve imaging parameters because they can affect image quality due to changes in the lighting and skin surface conditions of fish (Merz et al. 2012). These effects can obstruct the visualisation of the spot pattern in the area of interest, reducing the performance of the photo-identification method. Because the photography of this study was done under natural light but in different seasons of the year (i.e., spring for Calafquén Lake population and summer for Blue Back strain), lighting variations during photographing sessions cannot be ruled out. Therefore, the use of an artificial lighting system could be more appropriate in future studies, as this method has been shown to be useful for standardising lighting conditions during intensive photographing sessions related to the characterisation of rainbow trout skin pigmentation (Colihueque et al., 2011). However, the potential use of such a lighting system should be compact and portable in field applications. In juvenile Chinook salmon the occurrence of specular reflections in the captured images of the head spot patterns has been overcome by using fish submerged in water because this type of handling allows minimising the reflection of the skin surface (Merz et al. 2012). Therefore, this photographing method could be explored in future photo-identification studies of the rainbow trout to improve capture of the spot pattern on the flank. Of mention are the possible effects of other factors that might have affected the image quality of the analysed individuals, such as rearing conditions, sexual maturity, or stress, because in salmonids the available evidence indicates that these factors can cause the skin darkening (Colihueque 2010). For example, it has been observed that when rainbow trout individuals are maintained in tanks with a dark-coloured background, the overall skin darkening of them can increase significantly in the short-term (Green et al. 1991; Kasagi et al. 2020), as a consequence of physiological colour change, which could reduce the visualisation of the spot pattern. This phenomenon has been well supported by experimental observations in this species because individuals reared in darker tank colours are positively associated with darker body colours (Kasagi et al. 2020). In farmed populations, most of these factors can be controlled with appropriate management of farmed individuals before proceeding with photographing sessions; however, in wild populations, it is unlikely to be possible to control these factors. Nevertheless, currently the loss of image quality due to skin darkening can be overcome by digital image processing techniques by enhancing image contrast to extract the spot pattern before it is subjected to the recognition process, as verified by Merz et al. (2012) for juvenile Chinook salmon.

In Chile, several naturalised populations of rainbow trout inhabit different rivers and lakes (Colihueque et al. 2019). Thus, the present identification method may serve as a non-invasive tagging tool for the analysis of these populations. Specifically, this method could aid in answering certain biological questions related to demography, movement, and other ecological characteristics. For example, it could help address questions about the reproductive migratory behaviour of the naturalised rainbow trout living in southern Chile's Patagonian lakes during their spring run (Estay and Colihueque 2021); in this case, population monitoring throughout the maturation cycle is required as individuals move through various habitats that are available to them. Furthermore, it would be possible to use this identification technique while collecting various kind of information, such as information about growth, spawning dates, fecundity, and other reproductive variables, on specimens captured during different reproductive seasons; this would allow better understanding of their reproductive cycle. Furthermore, given the commercial significance of rainbow trout in our country (Colihueque and Estay 2010; Estay 2017), this non-invasive individual identification technique could be especially useful in hatcheries during controlled trials involving a small number of individuals to better understand reproduction, genetics, health, or nutrition, while avoiding the undesirable effects of traditional marking methods, such as PIT tagging, thereby enhancing animal welfare in aquaculture management. Increasing the effectiveness of the identification process with a fully automatic identification tool is another important issue, especially when a significant number of farm animals need to be examined. The future availability of such a tool would represent a step forwards for a long-term monitoring program based on low-cost and non-invasive techniques. We hope that our work will motivate other researchers to explore the skin pigmentation of rainbow trout for individual photo-identification.

5. Conclusion

The results of this study indicate high performance in the recognition of individuals based on the spot pattern after a relatively large number of individuals were analysed using software-driven processing. The long-term identification of rainbow trout individuals using this method also has favourable prospects because of its high recognition rate. Thus, this study demonstrates the utility of the flank skin spot patterns for photo-identification of rainbow trout individuals, given that this characteristic is unique for each individual. This photo-identification method opens the possibility to be used as an alternative non-invasive tagging method for addressing different biological questions related to this species, including issues that are relevant to fish farming. In trout farming this non-invasive individual identification technique could be especially useful in hatcheries during controlled trials involving a small number of individuals to better understand reproduction, genetics, health, or nutrition, while avoiding the undesirable effects of traditional marking methods, such as PIT tagging, thereby enhancing animal welfare in aquaculture management. This is the first study that explores the spot pattern of the flank skin of rainbow trout as a photo-identification method, but further analyses are needed to identify the different variables that may affect its performance.

6. Acknowledgements

We are indebted to Chiguaico Fish farm for the hatchery facilities and Lican Ray Fish and Game Club for its assistance in the field work during the collection of fish. We are grateful to the facilities provided by the fish farm Piscícola Huililco Ltda. for the realisation of this study.

7. References

Ade R (1989) Trout and salmon handbook. Facts on File, Inc., New York.

Aparicio E, García-Berthou E, Araguas RM, Martínez P, García-Marín JL (2005) Body pigmentation pattern to assess introgression by hatchery stocks in native *Salmo trutta* from Mediterranean streams. Journal of Fish Biology 67(4): 931–949. https://doi.org/10.1111/j.0022-1112.2005.00794.x

Arzoumanian Z, Holmberg J, Norman B (2005) An astronomical pattern-matching algorithm for computer-aided identification of whale sharks *Rhincodon typus*. Journal of Applied Ecology 42: 999–1011. https://doi.org/10.1111/j.1365-2664.2005.01117.x

Cisar P, Bekkozhayeva D, Movchan O, Saberioon M, Schraml R (2021) Computer vision based individual fish identification using skin dot pattern. Scientific Reports 11(1): 16904. https://doi.org/10.1038/s41598-021-96476-4

Clopper CJ, Pearson ES (1934) The use of confidence or fiducial limits illustrated in the case of the binomial. Biometrika 26: 404–413. https://doi.org/10.2307/2331986

Colihueque N (2010) Genetics of salmonid skin pigmentation: Clues and prospects for improving the external appearance of farmed salmonids. Reviews in Fish Biology and Fisheries 20(1): 71–86. https://doi.org/10.1007/s11160-009-9121-6

Colihueque N (2014) Analysis of the coloration and spottiness of Blue Back rainbow trout at a juvenile stage. Journal of Applied Animal Research 42(4): 474–480. https://doi.org/10.1080/09712119.2014.883317

Colihueque N, Parraguez M, Estay FJ, Diaz NF (2011) Skin color characterization in rainbow trout by use of computer-based image analysis. North American Journal of Aquaculture 73(3): 249–258. https://doi.org/10.1080/15222055.2011.581578

Colihueque N, Estay F (2010) Trout farming in Chile--country leads global production, further growth possible. Global Aquaculture Advocate February/J: 30–31.

Colihueque N, Estay FJ (2018) Association between the blue color intensity of the skin with growth performance and skin spottiness in a blue-colored strain of rainbow trout (*Oncorhynchus mykiss*). The Open Agriculture Journal 12(1): 251–261. https://doi.org/10.2174/1874331501819010251

Colihueque N, Estay FJ, Yáñez M (2018) Skin spotting variation associated with biometric and reproductive parameters in naturalized populations of rainbow trout, *Oncorhynchus mykiss*, from southern Chile. The Open Ecology Journal 11(1): 25–37. https://doi.org/10.2174/1874213001811010025

den Hartog J, Reijns R (2014) I3S Classic. Interactive individual identification system. Version 4.0.2.

Díaz NF, Estay FJ, Araneda C, Colihueque N (2011) Selección genética de rasgos del color de la piel de la trucha arcoiris *Oncorhynchus mykiss*. Versión Diferente Revista Salmón-Acuícola 8: 56–57.

Diez-Hermano S, Valero J, Rueda C, Ganfornina MD, Sanchez D (2015) An automated image analysis method to measure regularity in biological patterns: a case study in a *Drosophila* neurodegenerative model. Molecular Neurodegeneration 10(1): 9. https://doi.org/10.1186/s13024-015-0005-z

Estay F (2017) Una mirada a la industria de la trucha arcoíris cultivada en agua dulce en Chile y el mundo. Salmonexpert 7: 48–53.

Estay FJ, Colihueque N, Yáñez M (2021) Reproductive performance assessed during three spawning seasons in a naturalized rainbow trout population from southern Chile. Fisheries Research 244: 106107. https://doi.org/10.1016/j.fishres.2021.106107

Garcia de Leaniz C, Fraser N, Mikheev V, Huntingford F (1994) Individual recognition of juvenile salmonids using melanophore patterns. Journal of Fish Biology 45(3): 417–422. https://doi.org/10.1111/j.1095-8649.1994.tb01324.x

Green JA, Baker BI, Kawauchi H (1991) The effect of rearing rainbow trout on black or white backgrounds on their secretion of melanin-concentrating hormone and their sensitivity to stress. The Journal of endocrinology 128: 267–274. https://doi.org/10.1677/joe.0.1280267

Heumann C, Schomaker M, Shalabah (2016) Introduction to statistics and data analysis. Springer International Publishing AG, Cham, Switzerland, 456 pp. https://doi.org/10.1007/978-3-319-46162-5

Hillman G, Würsig B, Gailey G, Kehtarnavaz N, Drobyshevsky A, Araabi B, Tagare H, Weller D (2003) Computer-assisted photo-identification of individual marine vertebrates: A multi-species system. Aquatic Mammals 29: 117–123. https://doi.org/10.1578/016754203101023960

Hustler K (2014) Individual identification and visual indicators of breeding condition of banded kōkopu (*Galaxias fasciatus*) in an urban New Zealand stream. New Zealand Journal of Marine and Freshwater Research 48(1): 15–23. https://doi.org/10.1080/00288330.2013.826700

Islam M, Nose Y, Yasuda F (1973) Number and distribution pattern of the black spots on body surface of rainbow trout. Bulletin of the Japanese Society of Scientific Fisheries 39(7): 727–739.

Joo D, Kwan Y, Song J, Pinho C, Hey J, Won Y-J (2013) Identification of cichlid fishes from lake Malawi using computer vision. PLOS ONE 8(10): e77686. Available from: https://doi.org/10.1371/journal.pone.0077686.

Kasagi S, Miura M, Okazaki T, Mizusawa K, Takahashi A (2020) Effects of tank color brightness on the body color, somatic growth, and endocrine systems of rainbow trout *Oncorhynchus mykiss*. General and Comparative Endocrinology 298: 113581. https://doi.org/10.1016/j.ygcen.2020.113581

Katoh M, Tokimura M (2001) Genetic and morphological identification of *Sebastiscus tertius* in the East China Sea (Scorpaeniformes: Scorpaenidae). Ichthyological Research 48(3): 247–255. https://doi.org/10.1007/s10228-001-8142-5

Kause A, Ritola O, Paananen T, Eskelinen U, Mäntysaari E (2003) Big and beautiful? Quantitative genetic parameters for appearance of large rainbow trout. Journal of Fish Biology 62(3): 610–622. https://doi.org/10.1046/j.1095-8649.2003.00051.x

Lucentini L, Puletti ME, Ricciolini C, Gigliarelli L, Fontaneto D, Lanfaloni L, Bilò F, Natali M, Panara F (2011) Molecular and phenotypic evidence of a new species of genus *Esox* (Esocidae, Esociformes, Actinopterygii): The southern pike, Esox flaviae. PLOS ONE 6(12): e25218. https://doi.org/10.1371/journal.pone.0025218.

Martin-Smith KM (2011) Photo-identification of individual weedy seadragons *Phyllopteryx taeniolatus* and its application in estimating population dynamics. Journal of Fish Biology 78(6): 1757–1768. https://doi.org/10.1111/j.1095-8649.2011.02966.x

Meekan M, Bradshaw C, Press M, McLean C, Richards A, Quasnichka S, Taylor G (2006) Population size and structure of whale sharks *Rhincodon typus* at Ningaloo Reef, Western Australia. Marine Ecology Progress Series 319: 275–285. https://doi.org/10.3354/meps319275

Merz JE, Skvorc P, Sogard SM, Watry C, Blankenship SM, Van Nieuwenhuyse EE (2012) Onset of melanophore patterns in the head region of Chinook salmon: A natural marker for the reidentification of individual fish. North American Journal of Fisheries Management 32(4): 806–816. https://doi.org/10.1080/02755947.2012.681014

Parichy DM, Turner JM (2003) Zebrafish puma mutant decouples pigment pattern and somatic metamorphosis. Developmental Biology 256(2): 242–257. https://doi.org/10.1016/S0012-1606(03)00015-0

Pavlidis M, Papandroulakis N, Divanach P (2006) A method for the comparison of chromaticity parameters in fish skin: Preliminary results for coloration pattern of red skin Sparidae. Aquaculture 258(1): 211–219. https://doi.org/10.1016/j.aquaculture.2006.05.028

Persat H (1982) Photographic identification of individual grayling, *Thymallus thymallus*, based on the disposition of black dots and scales. Freshwater Biology 12(1): 97–101. https://doi.org/10.1111/j.1365-2427.1982.tb00606.x

Rodríguez FH, Cáceres G, Lhorente JP, Newman S, Bangera R, Tadich T, Neira R, Yáñez JM (2019) Genetic (co)variation in skin pigmentation patterns and growth in rainbow trout. Animal 13(4): 675–682. https://doi.org/10.1017/S175173111800188X

R Studio Team (2017) R Studio: Integrated development environment for R. RStudio, PBC, Boston, MA. http://www. Rstudio.com/.

Speed CW, Meekan MG, Bradshaw CJA (2007) Spot the match - wildlife photo-identification using information theory. Frontiers in zoology 4: 2. https://doi.org/10.1186/1742-9994-4-2

Stien LH, Nilsson J, Bui S, Fosseidengen JE, Kristiansen TS, Øverli Ø, Folkedal O (2017) Consistent melanophore spot patterns allow long-term individual recognition of Atlantic salmon. Journal of Fish Biology 91(6): 1699–1712. https://doi.org/10.1111/jfb.13491

Tack E (1973) Züchtungsfragen in der forellenzucht. Österreichs Fischerei 26(5/6): 77–84.

Table 1. Categories of spot pattern observed in two rainbow trout populations

			Categories					
		Body length (cm)	No dark spots		Well-formed dark spots		Diffused dark spots	
Population	n	(mean ± SD)	n	%	n	%	n	%
Calafquén Lake	56 (F=39, M=17)	49.1 ± 6.8	0	0.0	52	92.9	4	7.1
Blue Back strain	45 (F=40, M=5)	33.4 ± 3.0	4	8.9	39	86.7	2	4.4

n = total number of analysed fish, F = females, M = males, SD = standard deviation.

Table 2. Analysis of the spottiness regularity in the flank skin of rainbow trout

		No. of dark spots	CV-NN	Quartiles		
	n	Mean ± SD (range)	Mean ± SD (range)	Q1	Q2	Q3
Calafquén Lake	52	9.94 ± 4.04 (2-21)	0.58 ± 0.92 (0.00-5.27)	0.23	0.33	0.43
Blue Back strain	39	17.46 ± 9.20 (3-54)	0.73 ± 1.38 (0.00-6.84)	0.27	0.33	0.50
Regularity control	29	9.76 ± 1.24 (9-13)	0.12 ± 0.16 (0.02-0.40)	0.02	0.02	0.33

n = total number of analysed fish, CV-NN = coefficients of variation for nearest-neighbour distances between spots, SD = standard deviation.

Table 3. Ranking score image for recognition of rainbow trout individuals based on the flank spot pattern

			Ranking score image		
		No. of dark spots	1st ranked	2nd ranked	3rd ranked
	n	Mean ± SD (range)	Mean ± SD (range)	Mean ± SD (range)	Mean ± SD (range)
Calafquén Lake	52	57.19 ± 20.26 (16-97)	0 ± 0 (0-0)	7.02 ± 4.34 (3.28-27.29)	7.76 ± 4.53 (3.67-28.40)
Blue Back strain	39	60.82 ± 26.36 (12-96)	0 ± 0 (0-0)	8.28 ± 11.17 (2.39-49.15)	9.16 ± 12.84 (2.67-55.81)

n = total number of analysed fish, SD = standard deviation.

Table 4. Ranking score image for long-term recognition of rainbow trout individuals evaluated at two consecutive photographing sessions

	Year 2007 (2-year-old)								
PIT tag	0006-A2B7E9	0006-A2E845	0006-A2BA29	0006-A2B748	0006-A2CB95	0006-A2EG41	0006-A2C072	0006-A2CF16	Mean ± SD
No. of spots	44	48	73	12	40	49	80	84	53.75 ± 24.08 [a]
Body length (cm)	32.2	30.3	34.6	32.7	30.4	30.7	33.7	33.2	32.23 ± 1.62 [x]
	Year 2008 (3-year-old)								
PIT tag	0006-A2B7E9	0006-A2E845	0006-A2BA29	0006-A2B748	0006-A2CB95	0006-A2EG41	0006-A2C072	0006-A2CF16	Mean ± SD
No. of spots	65	79	65	19	40	45	74	69	57.00 ± 20.43 [a]
Body length (cm)	45.5	41.5	40.5	44.4	47.3	49.1	38.9	40.5	43.46 ± 3.66 [y]
	Ranking score image								
Ranking 1	**4.14**	**1.61**	**2.38**	**40.48**	5.44	**1.69**	**1.47**	2.86	7.51 ± 13.39
Ranking 2	4.77	3.31	2.76	65.27	6.91	3.42	2.47	3.22	11.52 ± 21.77
Ranking 3	6.11	3.37	2.77	72.74	**7.22**	3.70	2.58	3.28	12.72 ± 24.31
Ranking 4	6.19	3.95	3.34	73.69	8.55	4.22	2.94	**3.41**	13.29 ± 24.48
Ranking 5	6.55	4.21	3.75	82.14	8.69	4.36	2.95	3.98	14.58 ± 27.36
Ranking 6	6.83	4.41	4.20	135.87	9.09	5.27	4.50	4.99	21.90 ± 46.08
Ranking 7	7.87	6.09	7.44	135.94	9.86	5.85	7.20	7.11	23.42 ± 45.48
Ranking 8	28.58	24.77	25.24	183.22	13.69	26.58	30.09	25.04	44.65 ± 56.21

Different letters for the number of spots and body length indicate a significant differences between means according to Student's t-test ($P < 0.05$). Bold indicates correct match values, SD = standard deviation.

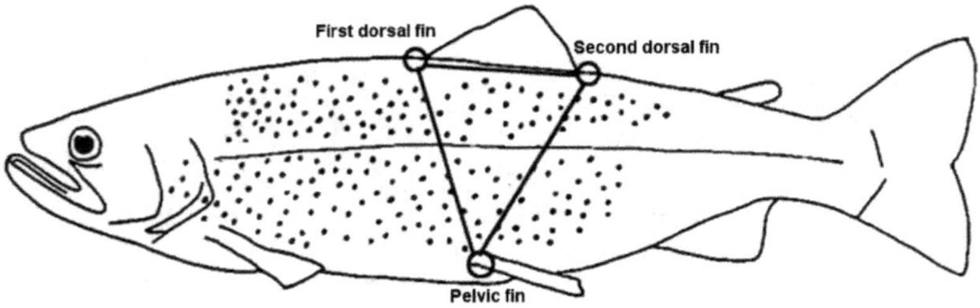

Fig.1 Region of interest used in the photo-identification of rainbow trout individuals. The region of interest was defined by a triangle bounded by three fixed anatomical landmarks that included the initial dorsal fin (first dorsal fin), the end dorsal fin (second dorsal fin), and the pelvic fin.

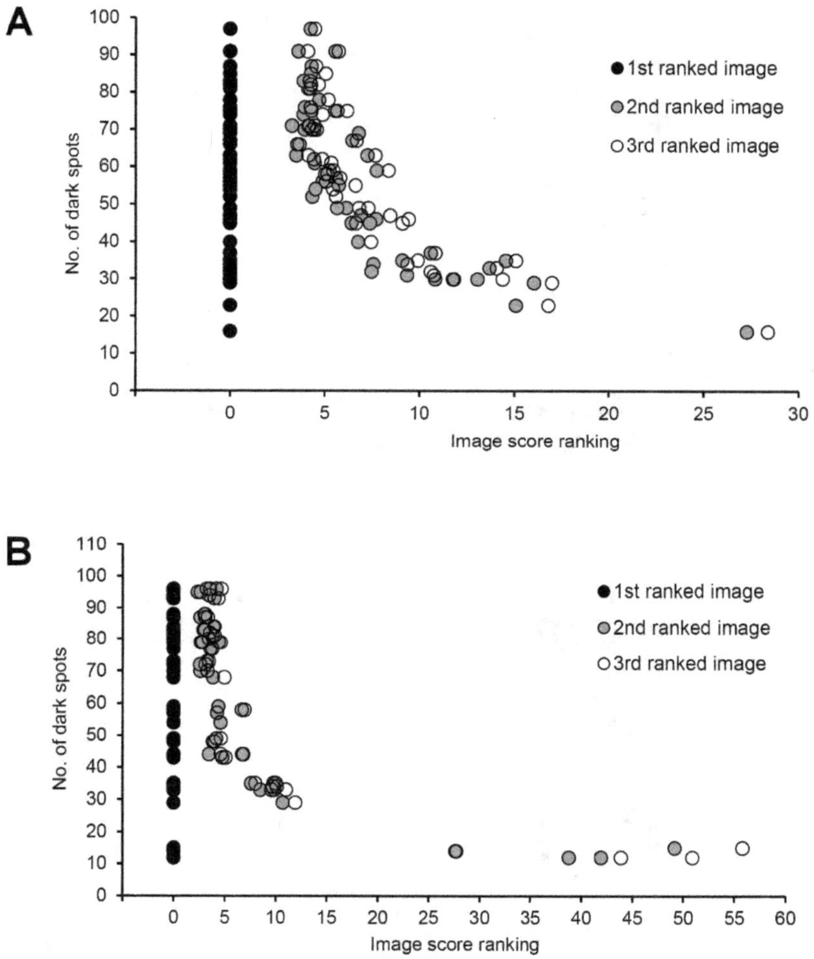

Fig. 2 Ranking score of all annotated images when they were compared with the annotated images stored in the identification database. Calafquén Lake population (A), Blue Back strain (B). Annotated images that achieved the first ranking score (ranking score = 0) represent the perfect recognition of the same individual.

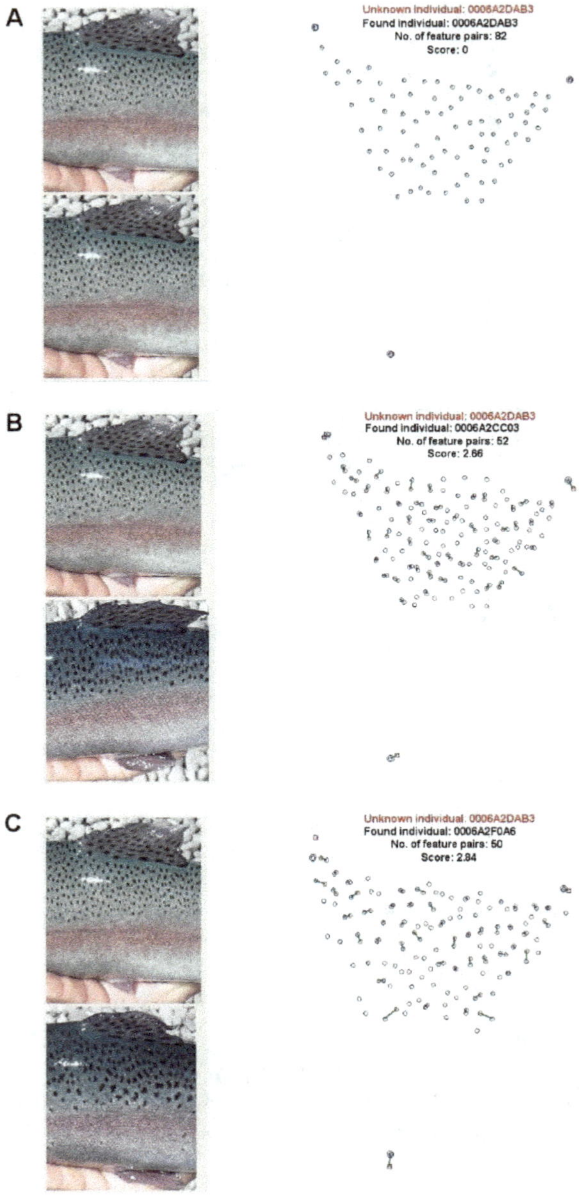

***Fig. 3** An example of the perfect recognition of an individual of the Blue Back strain. First ranked image (A), second ranked image (B), and third ranked image (C). A ranking score equal to zero (first ranked image) indicates the perfect recognition of the same individual, indicating a similar spot position between the two point clouds. The annotated image was compared with all annotated images stored in the identification database using the I3S CLASSIC v.4.02 software.*

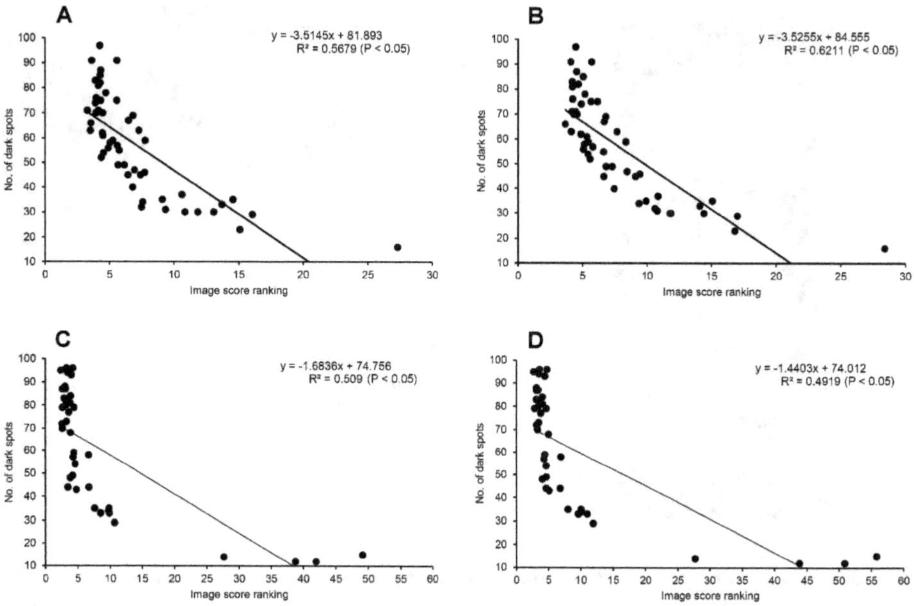

Fig. 4 Correlations between the number of spots and the ranking score for second- and third-ranked annotated images. Second-ranked images (A) and third-ranked images (B) of Calafquén Lake population; second-ranked images (C), third-ranked images (D) of Blue Back strain. The regression line is indicated in a black line.

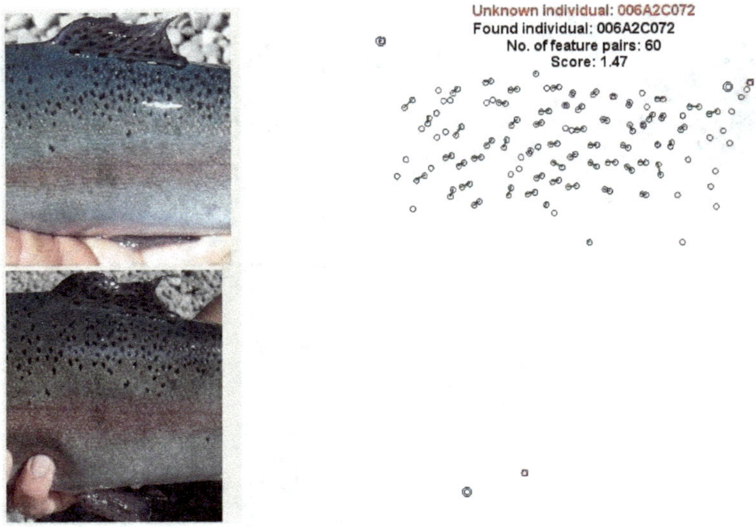

Fig. 5 A long-term photo-identification of rainbow trout individuals. A subset of images from the Blue Back strain was selected to determine if the flank skin spots present at 2-years-old were still present at 3-years-old. The first image ranking, as shown in this figure, indicates a complete match of the spot pattern between the above and below images, indicating the perfect recognition of the individual identified with PIT tag 0006A2C072. The I3S CLASSIC v.4.02 software was used to compare an individual's spot patterns at 2-years-old with spot patterns from multiple individuals at 3-years-old.

www.ingramcontent.com/pod-product-compliance
Lightning Source LLC
LaVergne TN
LVHW022304200926
848481LV00004B/1209
* 9 7 9 8 8 9 9 6 6 6 8 1 0 *